The Concise Geologic Time Scale

This concise handbook presents a summary of Earth's history over the past 4.5 billion years as well as a brief overview of contemporaneous events on the Moon, Mars, and Venus. The authors have been at the forefront of chronostratigraphic research and initiatives to create an international geologic time scale for many years, and the charts in this book present the most up-to-date, international standard, as ratified by the International Commission on Stratigraphy and the International Union of Geological Sciences. This book is an essential reference for all geoscientists, including researchers, students, and petroleum and mining professionals. The presentation is non-technical and illustrated with numerous color charts, maps, and photographs. The book also includes a laminated card of the complete time scale for use as a handy reference in the office, laboratory, or field.

JAMES OGG is a Professor in the Department of Earth and Atmospheric Sciences at Purdue University and has served as Secretary-General of the International Commission on Stratigraphy since 2000. As part of this role, he developed the TimeScale Creator databases and visualization system (freely available at www.stratigraphy.org). His research specialties include Mesozoic marine stratigraphy, paleomagnetism, and climate cycles.

GABI OGG is a micropaleontologist and is responsible for the many time scale charts and other graphics in this book and numerous other publications.

FELIX GRADSTEIN is Professor of Stratigraphy and Micropaleontology at the Geology Department of the Natural History Museum of Oslo University. He was chair of the International Commission on Stratigraphy from 2000 to 2008, and under his tenure major progress was made with the definition and ratification and international acceptance of chronostratigraphic units from Precambrian through to Quaternary.

GEOLOGIC TIME SCALE

PHANEROZOIC | PRECAMBRIAN

CENOZOIC

AGE (Ma)	Period	Epoch	Age / Stage	AGE (Ma)
0	Quaternary*	Pleisto-cene	"Ionian"	
			Calabrian	0.78
			Gelasian	1.81
		Pliocene	Piacenzian	2.59
			Zanclean	3.60
5				5.33
	Neogene	Miocene (L)	Messinian	7.25
			Tortonian	
10				11.61
		Miocene (M)	Serravallian	13.82
15			Langhian	15.97
		Miocene (E)	Burdigalian	
20				20.43
			Aquitanian	23.03
25	Paleogene	Oligocene (L)	Chattian	
				28.4
30		Oligocene (E)	Rupelian	
35				33.9
		Eocene (L)	Priabonian	37.2
40			Bartonian	40.4
		Eocene (M)	Lutetian	
45				
				48.6
50		Eocene (E)	Ypresian	
55				55.8
		Paleocene (L)	Thanetian	58.7
60		Paleocene (M)	Selandian	61.1
		Paleocene (E)	Danian	
65				65.5

MESOZOIC

AGE (Ma)	Period	Epoch	Age / Stage	AGE (Ma)
65	Cretaceous	Late	Maastrichtian	65.5
70				70.6
75			Campanian	
80				
85			Santonian	83.5 / 85.8
			Coniacian	88.6
90			Turonian	93.6
95			Cenomanian	
100		Early		99.6
105			Albian	
110				112.0
115			Aptian	
120				
125			Barremian	125.0
130			Hauterivian	130.0
135			Valanginian	133.9
140			Berriasian	140.2
145				145.5
150	Jurassic	Late	Tithonian	150.8
155			Kimmeridgian	155.6
160			Oxfordian	161.2
165		Middle	Callovian	164.7
			Bathonian	167.7
170			Bajocian	171.6
175			Aalenian	175.6
180		Early	Toarcian	183.0
185			Pliensbachian	
190				189.6
195			Sinemurian	196.5
200			Hettangian	199.6
	Triassic	Late	Rhaetian	203.6
205				
210			Norian	
215				216.5
220			Carnian	
225				228.7
230		Middle	Ladinian	
235				237.0
240			Anisian	
245		Early	Olenekian	245.9
250			Induan	249.5 / 251.0

PALEOZOIC

AGE (Ma)	Period	Epoch	Age / Stage	AGE (Ma)
255	Permian	Lopingian	Changhsingian	251.0 / 253.8
260			Wuchiapingian	260.4
265		Guada-lupian	Capitanian	265.8
270			Wordian	268.0
			Roadian	270.6
275		Cisuralian	Kungurian	275.6
280			Artinskian	
285				284.4
290			Sakmarian	
295			Asselian	294.6
300	Carboniferous	Pennsylvanian (Late)	Gzhelian	299.0
305			Kasimovian	303.4
310		Pennsylvanian (Middle)	Moscovian	307.2
315		Pennsylvanian (Early)	Bashkirian	311.7
320				318.1
325		Mississippian (Late)	Serpukhovian	
330				328.3
335		Mississippian (Middle)	Visean	
340				
345				345.3
350		Mississippian (Early)	Tournaisian	
355				
360				359.2
365	Devonian	Late	Famennian	
370				374.5
375			Frasnian	
380				385.3
385		Middle	Givetian	
390			Eifelian	391.8
395				397.5
400		Early	Emsian	
405				407.0
410			Pragian	411.2
415			Lochkovian	416.0
420	Silurian	Pridoli		418.7
		Ludlow	Ludfordian	421.3
			Gorstian	422.9
425		Wenlock	Homerian	426.2
			Sheinwoodian	428.2
430		Llandovery	Telychian	
435				436.0
440			Aeronian	439.0
			Rhuddanian	443.7
445			Hirnantian	445.6
450	Ordovician	Late	Katian	
455				455.8
460			Sandbian	460.9
465		Middle	Darriwilian	468.1
470			Dapingian	471.8
475		Early	Floian	478.6
480				
485			Tremadocian	
490	Cambrian	Furongian	Stage 10	488.3 / 492
495			Stage 9	496
500			Paibian	499
505		Epoch 3	Guzhangian	503
			Drumian	506.5
510			Stage 5	510
515		Epoch 2	Stage 4	515
520			Stage 3	521
525		Terreneuvian	Stage 2	528
530				
535			Fortunian	
540				542.0

PRECAMBRIAN

AGE (Ma)	Eon	Era	Period	AGE (Ma)
600	Proterozoic	Neoproterozoic	Ediacaran	542
700			Cryogenian	~635
800				850
900			Tonian	
1000				1000
1100		Mesoproterozoic	Stenian	
1200				1200
1300			Ectasian	
1400				1400
1500			Calymmian	
1600				1600
1700		Paleoproterozoic	Statherian	
1800				1800
1900			Orosirian	
2000				2050
2100			Rhyacian	
2200				2300
2300			Siderian	
2400				
2500	Archean	Neoarchean		2500
2600				
2700				
2800				2800
2900		Mesoarchean		
3000				
3100				
3200		Paleoarchean		3200
3300				
3400				
3500				
3600		Eoarchean		3600
3700				
3800				
3900				
4000		Hadean		
4100				
4200				
4300				
4400				
4500				
4600				

* Definition of the Quaternary and revision of the Pleistocene are under discussion. Base of the Pleistocene is at 1.81 Ma (base of Calabrian), but may be extended to 2.59 Ma (base of Gelasian). The historic "Tertiary" comprises the Paleogene and Neogene, and has no official rank.

The Concise
Geologic
Time Scale

James G. Ogg
Purdue University, Indiana

Gabi Ogg

and

Felix M. Gradstein
University of Oslo

CAMBRIDGE
UNIVERSITY PRESS

CAMBRIDGE UNIVERSITY PRESS
Cambridge, New York, Melbourne, Madrid, Cape Town, Singapore, São Paulo, Delhi, Dubai, Tokyo

Cambridge University Press
The Edinburgh Building, Cambridge CB2 8RU, UK

Published in the United States of America by Cambridge University Press, New York

www.cambridge.org
Information on this title: www.cambridge.org/9780521898492

First published 2008
Reprinted 2010

Printed in the United Kingdom at the University Press, Cambridge

A catalog record for this publication is available from the British Library

Library of Congress Cataloging in Publication data
Ogg, James G. (James George), 1952–
Concise geologic time scale / James G. Ogg, Gabi Ogg, and Felix M. Gradstein.
    p.   cm.
ISBN 978-0-521-89849-2
1. Geological time.   2. Geology, Stratigraphic.   I. Ogg, Gabi.   II. Gradstein, F. M.   III. Title.
QE508.O34   2008
551.7–dc22

                                                                    2008025656

ISBN 978-0-521-89849-2 hardback

Contents

Of the many advances in chronostratigraphy and other fields of stratigraphy, we wish to highlight two major updates to the International Geologic Time Scale that have occurred after May, 2008:

(1) The **Quaternary** has been ratified as a System/Period, with its base at the Gelasian GSSP (ca. 2.6 Ma) by IUGS on 29 June 2009. The Gelasian Stage/Age was transferred to the Pleistocene Series/Epoch. The underlying System/Period is the Neogene.

(2) The **base-Jurassic** GSSP was ratified as the Kuhjoch section (Tyrol, Austria) to coincide with the first appearance of *Psiloceras spelae tirolicum* (first species of this *Psiloceras* group). This level is estimated to be approximately 100 kyr later than the negative carbon isotope excursion that is attributed to the Central Atlantic magmatic province (ca. 201.5 Ma).

Current GSSP definitions, graphics and other information are posted at the website of the Subcommission for Stratigraphic Information of the International Commission on Stratigraphy (ICS): *http://stratigraphy.science.purdue.edu.*

1
Introduction

This book

The geologic time scale is the framework for deciphering the history of our planet Earth.

This book is a summary of the status of that scale and some of the most common means for global correlation. It is intended to be a handbook; therefore, readers who desire more background or details on any aspect should utilize the suggested references at the end of each section, especially the detailed compilations in *A Geologic Time Scale 2004* (GTSO4).

Each chapter spans a single period/ system, and includes:

(1) International divisions of geologic time and their global boundaries.

(2) Selected biologic, chemical, sea-level, geomagnetic and other events or zones.

(3) Estimated numerical ages for these boundaries and events.

(4) Selected references and websites for additional information on each period.

We are constantly improving and enhancing our knowledge of Earth history, and simultaneously attaining a global standardization of nomenclature. Therefore, any geologic time scale represents a status report in this grand endeavor. The international divisions in this document represent the decisions and recommendations of the International Commission on Stratigraphy (ICS), as ratified by the International Union of Geological Sciences (IUGS) through March 2008, plus proposed or working definitions for the remaining geologic stages. For consistency and clarity, it was decided to retain the same numerical time scale that was used in *A Geologic Time Scale 2004* (Gradstein *et al.*, 2004) for the majority of the stage boundaries, except if the ratified definitions after 2004 for those boundaries are at a different level from the previous "working" versions (e.g., base of Serravallian). We have made an effort, where applicable, to partially update and enhance the events of the selected biologic, chemical, and sea-level columns and their relative scaling within each stage using accepted or proposed calibrations through October 2007. These stratigraphic scales are a small subset of the compilations in *TimeScale Creator*, a public database visualization system available through the ICS website (*www.stratigraphy.org*). This ICS website should also be visited for the updated charts on international subdivisions, status of boundary decisions, and other time-scale-related information.

International divisions of geologic time and their global boundaries

One must have a common and precise language of geologic time to discuss and unravel Earth's history. One of the main goals of the International Commission on Stratigraphy and its predecessors under the International Geological Congresses (IGC) has been to unite the individual regional scales by reaching agreement on a standardized nomenclature and hierarchy for stages defined by precise Global Boundary Stratotype Sections and Points (GSSPs).

The choice of an appropriate boundary is of paramount importance. "*Before formally defining a geochronologic boundary by a GSSP, its practical value – i.e., its correlation potential – has to be thoroughly tested. In this sense, correlation preceded definition.*" (Remane,

2003). "*Without correlation, successions in time derived in one area are unique and contribute nothing to understanding Earth history elsewhere.*" (McLaren, 1978). Most GSSPs coincide with a single primary marker, which is generally a biostratigraphic event, but other stratigraphic events with widespread correlation should coincide or bracket the GSSP. Other criteria include avoidance of obvious hiatuses near the boundary interval and accessibility (see Table 1.1).

This task proved to be more challenging than envisioned when the GSSP effort began in the 1980s. The choice of the primary criteria for an international stage boundary can be a contentious issue, especially when competing regional systems or vague historical precedents are involved. Preference for stratigraphic priority is laudable when selecting GSSPs, but subsidiary to scientific and practical merit if the historical versions are unable to provide useful global correlations. Therefore, the Cambrian and the Ordovician subcommissions developed a global suite of stages that have demonstrated correlation among regions, in contrast to any of the American, British, Chinese, or Australian regional suites. However, such regional stages are very useful; and this book presents selected inter-regional correlation charts as appropriate.

Approximately one-third of the 100 geologic stages await international definition with precise GSSPs. Those that remain undefined by boundary definitions have either encountered unforeseen problems in recognizing a useful correlation horizon for global usage (e.g., base of Cretaceous System), a desire to achieve

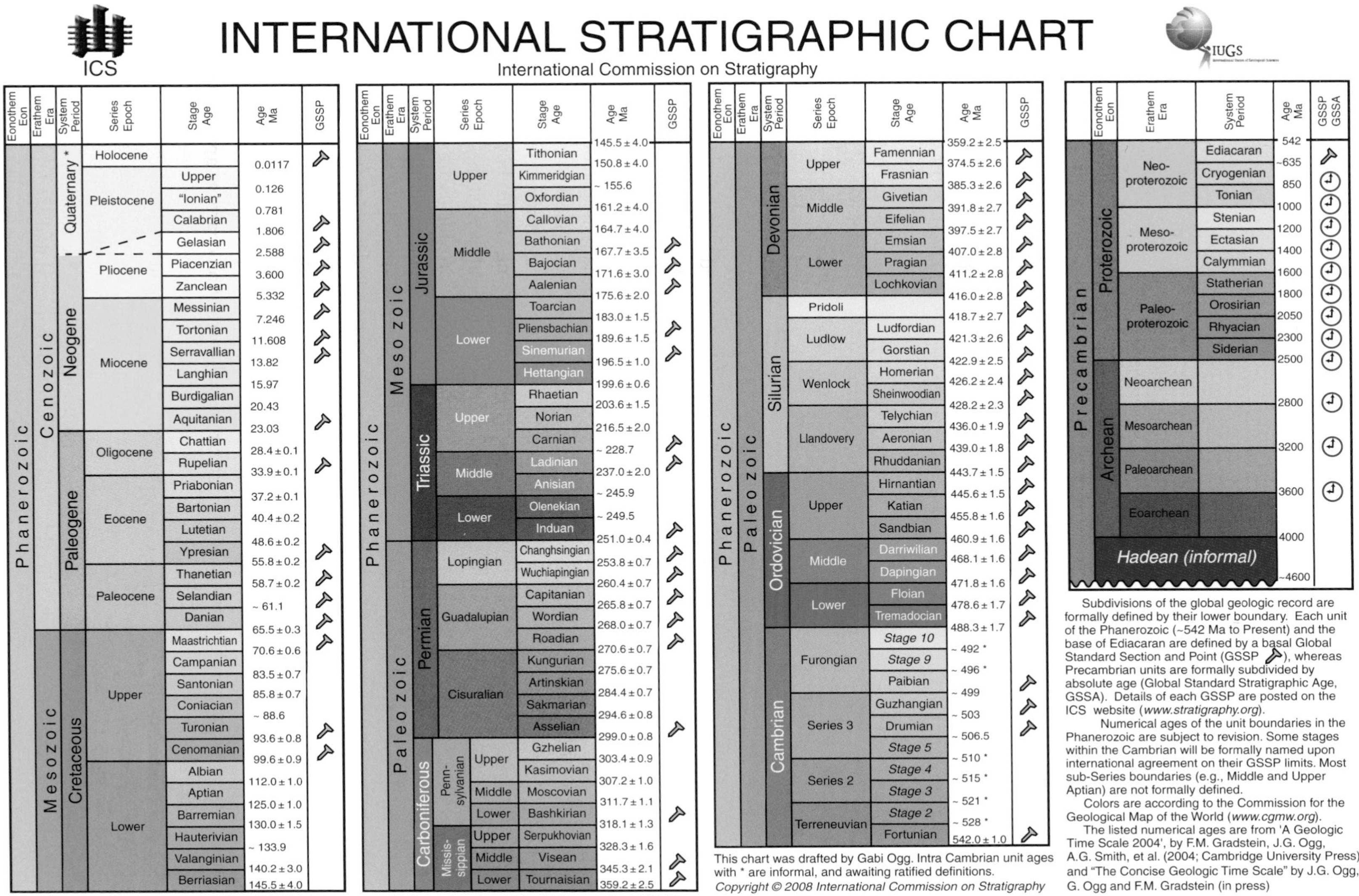

INTERNATIONAL STRATIGRAPHIC CHART

International Commission on Stratigraphy

Figure 1.1. International divisions of geologic time and ratified GSSPs (status as of March, 2008).

Table 1.1 Requirements for establishing a Global Stratotype Section and Point (GSSP)
(1) Name and stratigraphic rank of the boundary Including concise statement of GSSP definition
(2) GSSP geographic and physical geology Geographic location, including map coordinates Geologic setting (lithostratigraphy, sedimentology, paleobathymetry, post-depositional tectonics, etc.) Precise location and stratigraphic position of GSSP level and specific point Stratigraphic completeness across the GSSP level Adequate thickness and stratigraphic extent of section above and below Accessibility, including logistics, national politics and property rights Provisions for conservation and protection
(3) Primary and secondary markers Principal correlation event (marker) at GSSP level Other primary and secondary markers – biostratigraphy, magnetostratigraphy, chemical stratigraphy, sequence stratigraphy, cycle stratigraphy, other event stratigraphy, marine–land correlation potential Potential age dating from volcanic ashes and/or orbital tuning Demonstration of regional and global correlation
(4) Summary of selection process Relation of the GSSP to historical usage; references to historical background and adjacent (stage) units; selected publications Other candidates and reasons for rejection; summary of votes and received comments Other useful reference sections
(5) Official publication Summary for full documentation in IUGS journal *Episodes* Digital stratigraphy (litho-, paleo-, magneto-, chemical) images and graphic files submitted to ICS for public archive Full publication (if not in *Episodes*) in an appropriate journal

Source: Revised from Remane *et al.* (1996) according to current procedures and recommendations of the IUGS's International Commission on Stratigraphy (ICS). Modified from Figure 2.2 in *A Geologic Time Scale 2004*.

calibration to other high-resolution scales (e.g., base of Langhian Stage in Miocene awaiting astronomical tuning), inability to reach majority agreement, or other difficulty. In these cases, this book presents the status or temporary working definition of the yet-to-be-defined stages/ages within each system/period. One unresolved GSSP is the base of Quaternary for which the IUGS–IGC has not yet ratified a standardized definition or rank.

Geologic time and the observed rock record are separate but related concepts. A geologic time unit (geochronologic unit) is an abstract concept measured from the rock record by radioactive decay, Milankovitch cycles or other means. A "rock-time" or chronostratigraphic unit consists of the total rocks formed globally during a specified interval of geologic time. Therefore, a parallel

nomenclature system has been codified – geologic-time units of period/epoch/age that span the rock-record units of system/series/stage. The period/systems are grouped into eras/erathems within eons/eonthems, respectively. [The usage of the term "age" as the time-unit spanning the rock-unit of "stage" has received criticism from geochronologists, and was omitted "to avoid some ambiguity and confusion" in some time-scale publications (e.g., Harland *et al.*, 1982, 1989; Gradstein *et al.*, 2004). In this version, the age/stage duality is denoted in figure captions.] The same philosophy applies to successions, in which the terms of "Early/Late" are used when discussing time events and for the formal names of epochs on time scales, whereas "Lower/Upper" are used on stratigraphic columns and for formal names of series.

Biologic, chemical, sea-level, geomagnetic and other events or zones

Geologic stages are recognized, not by their boundaries, but by their content. The rich fossil record remains the main method to distinguish and correlate strata among regions, because the morphology of each taxon is the most unambiguous way to assign a relative age. The evolutionary successions and assemblages of each fossil group are generally grouped into zones. We have included selected zonations and/or events (first or last appearance datum, FAD or LAD) for widely used biostratigraphic groups in each system/period.

Trends and excursions in stable-isotope ratios, especially of carbon 12/13 and strontium 86/87, have become an increasingly reliable method to correlate among regions. Some of the carbon-isotope excursions are associated with widespread deposition of organic-rich sediments. Ratios of oxygen 16/18 are particularly useful for the glacial–interglacial cycles of Pliocene–Pleistocene. Sea-level trends, especially rapid oscillations that caused widespread exposure or drowning of coastal margins, can be associated with these isotopic-ratio excursions; but the synchroneity and driving cause of pre-Neogene sequences is disputed. We have included major sequences as interpreted by widely used publications, but many of these remain to be documented as global eustatic sea-level oscillations.

Geomagnetic polarity chrons are well established for correlation of marine magnetic anomalies of latest Jurassic through Holocene to the magnetostratigraphy of fossiliferous strata.

Pre-Kimmeridgian magnetic polarity chrons have been verified in some intervals, but exact correlation to biostratigraphic zonations remains uncertain for many of these. The geomagnetic scales on diagrams in this book are partly an update of those compiled for GTS04.

Methods for assigning numerical ages

The Quaternary–Neogene is the only interval in which high-resolution ages can be assigned to most biostratigraphic, geomagnetic and other events, including stage GSSPs. In the majority of this upper Cenozoic, especially for the interval younger than about 14 myr (millions of years), series of investigations have compiled the record of climatic–oceanic changes associated with periodic oscillations in the Earth's orbital parameters of precession, obliquity, and eccentricity as derived from astronomical models of the Solar System. This astronomical-tuned time scale will soon be extended to the currently "floating" orbital durations of Paleogene strata and into the Cretaceous. Orbital-cycle ("Milankovitch") durations have been achieved for portions of older periods (e.g., geomagnetic scale for Late Triassic); but the calibration of these intervals to numerical ages depends upon constraints from radiometric ages.

Dates derived from radioisotopic methods on minerals in volcanic ashes interbedded with fossiliferous sediment provide a succession of constraints on estimating

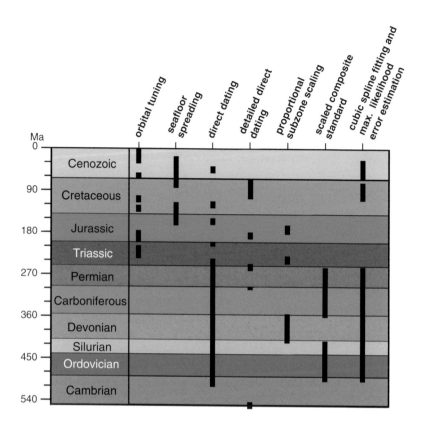

Figure 1.2. Methods used to construct *A Geologic Time Scale 2004* (GTS04) integrated different techniques depending on the quality of data available within each interval.

numerical ages for the geologic time scale. These methods and discussion of uncertainties are summarized in *A Geologic Time Scale 2004* and other publications. The ages of events and stage boundaries that are between the selected radiometric dates are interpolated according to their relative position in composite sediment sections (constrained optimization or graphical correlation procedures), their relative correlation to a smoothed scale of marine magnetic anomalies, their level within an orbital-cycle-scaled succession, or less quantitative means. A goal of geochronologists and database compilers is to progressively narrow the uncertainties on such interpolations and converge on exact numerical ages for all events.

For clarity, the numerical age is abbreviated as "a" (for annum), "ka" for thousands, "Ma" for millions, and "Ga" for billions of years before present. The elapsed time or duration is abbreviated as "yr" (for year), "kyr" (thousands of years), or "myr" (millions of years). Ages are given in years before "Present" (BP). To avoid a constantly changing datum, "Present" had been fixed as AD 1950 (as in carbon-14 determinations), the date of the beginning of modern isotope dating research in laboratories around the world, but the confusing offset between the current year and "Present" has led many Holocene workers to use a "BP2000," which is relative to the year AD 2000.

It has been suggested that the same unit should be used for absolute and relative measurements in time; therefore, elapsed time or duration should also be abbreviated as ka or Ma. This is similar to the use of K for both actual

Table 1.2 Modified ages of stage boundaries in this book relative to *A Geologic Time Scale 2004.*

Chronostratigraphic unit	Age in GTS04	Age in this book	Summary
base Holocene Epoch/Series (in Quaternary)	11.5ka	**11.7ka**	Holocene GSSA (based on GSSP in ice cores) is assigned as 11 700yr before AD 2000
base Serravallian Stage (in Neogene)	13.65Ma	**13.82Ma**	Ratified GSSP coincides with the end of major Mi-3b cooling step in oxygen isotopes; about 0.17myr earlier than the GTS04 provisional version
base Selandian Stage (in Paleogene)	61.7Ma	**61.1Ma**	Pending GSSP is at the onset of a carbon-isotope shift; about 0.6myr younger than GTS004 provisional version
base Coniacian Stage (in Cretaceous)	89.3Ma	**88.6Ma**	Inoceramid marker for base Coniacian is now known to be significantly younger than the GTS04 correlation to the base of the *F. petrocoriensis* ammonite zone
base Hauterivian Stage (in Cretaceous)	136.4Ma	**133.9Ma**	Hauterivian/Valanginian boundary is now known to be near the base of polarity Chron M10n (not M11n), which is consistent with the cycle-scaled duration of the Hauterivian and Valanginian
base Kimmeridgian Stage (in Jurassic)	155.7Ma	**155.6Ma**	Assignment to nearest 0.1 myr relative to M-sequence marine magnetic model
base Carnian Stage (in Triassic)	228.0Ma	**228.7Ma**	Carnian now includes the *D. canadensis* ammonite zone, which was placed in Anisian in GTS04
base Anisian Stage (in Triassic)	245.0Ma	**245.9Ma**	Base of Anisian is tentatively shifted down by one conodont zone; assigned age is from cycle-mag scaling relative to base Triassic
base Olenekian Stage (in Triassic)	249.7Ma	**249.5Ma**	Base of Olenekian shifted downward by one brief ammonite zone; assigned age is from cycle-mag scaling relative to base Triassic
base Gzhelian Stage (in Carboniferous)	303.9Ma	**303.4Ma**	The basal Gzhelian conodont marker has a revised correlation to fusulinid zones, and is cycle-scaled relative to base Permian
base Kasimovian Stage, base Upper Pennsylvanian Series (in Carboniferous)	306.5Ma	**307.2Ma**	Kasimovian is essentially shifted by one fusulinid zone (current preferred); but there is no final decision; age is cycle-scaled relative to base Permian
base Serpukhovian, base Upper Mississippian Series (in Carboniferous)	326.4Ma	**328.3Ma**	Proposed marker for base of Serpukhovian is now the base of conodont zone of *Lochriea ziegleri*; GTS04 definition was ~2 myr higher at base of conodont zone *Lochriea cruciformis*
base Stage 10 of Cambrian		**~492Ma**	New Cambrian stages/series and scaling of zones, with ages estimated by Peng and Babcock (Chapter 4)
base Stage 9 of Cambrian		**~496Ma**	As above
base Paibian Stage, base Furongian Series (in Cambrian)	501.0	**~499Ma**	As above
base Guzhangian Stage (in Cambrian)		**~503Ma**	As above
base Drumian Stage (in Cambrian)		**~506.5Ma**	As above
base Stage 5 and base Series 3 of Cambrian		**~510Ma**	As above
base Stage 4 of Cambrian		**~515Ma**	As above
base Stage 3 and base Series 2 of Cambrian		**~521Ma**	As above
base Stage 2 of Terreneuvian Series of Cambrian		**~528Ma**	As above
base Ediacaran System (in Neoproterozoic)	630Ma	**635Ma**	Radiometric ages revise the previous rounded estimate

temperature and a temperature difference. In such a system with a single unit, the Aptian begins at "125 Ma" and spans "13 Ma." However, usage for time units in geosciences are far from standardized among scientific journals and organizations; and to avoid any confusion, we will continue the dichotomy of Ma/myr for age/duration.

In the years after the computation of the numerical scales in GTS04, major advances have occurred in radiometric dating, including: (1) improved analytical procedures for obtaining uranium–lead ages from zircons that shifted published ages for some levels by more than 1 myr, (2) an astronomically dated neutron irradiation monitor for ^{40}Ar–^{39}Ar methods implying earlier reported ages should be shifted older by nearly 1%, (3) technological advances that reduce uncertainties and enabled acquisition of reduced-error results of the rhenium-187 to osmium-187 (Os–Re) chronometer in organic-rich sediments [e.g., 154.1±2.2 Ma on the proposed base-Kimmeridgian GSSP (Selby, 2007)], and (4) the continued acquisition of additional radiometric ages. These exciting advances have led to several suggestions for revision of assigned or interpolated ages for geologic stages and component events. In each chapter of this book, we indicate how some of these new results and methods may modify the estimated numerical scales, but have not attempted to make a new set of numerical scales. Such a comprehensive revision is being compiled by the different groups for the enhanced GTS2010 book (see below).

In this book, we have retained the assigned ages for stage boundaries in GTS04, but have greatly improved the scaling and correlations of different biostratigraphic events and other stratigraphic information that are within those stages. However, it was necessary to update some of the stage boundaries (Table 1.2). These revisions mainly reflect decisions on GSSPs or potential GSSP markers, which had been given tentative working definitions in GTS04, and on the establishment of the stage/series framework for the Cambrian. Except in certain cases (Early Triassic, Late Carboniferous), the "primary age scales" that were calculated in GTS04 (C-sequence and M-sequence chrons, ammonite zones, graphical composite standard for Carboniferous, CONOP composite for Ordovician–Silurian graptolites, etc.) have been retained for assigning ages to most other events in this book. However, advances in cycle stratigraphy, additional radiometric dates, revised standards and methods of processing radiometric samples, and new interpreted correlations imply that portions of these reference time scales will require significant modification in the future (see discussions in each chapter).

TimeScale Creator database and chart-making package

One goal of ICS is to provide detailed global and regional "reference" scales of Earth history. Such scales summarize our current consensus on the

Figure 1.3. Age calibration for *A Geologic Time Scale 2004*. The precision of individual radiometric dates and the final inferred precision on stage boundaries (red line) plotted in terms of precision (%) instead of absolute uncertainty (in myr). Radiometric age dates published after GTS2004 have confirmed the interpolated geochronology of the Late Jurassic-Early Cretaceous, thereby reduce the uncertainty.

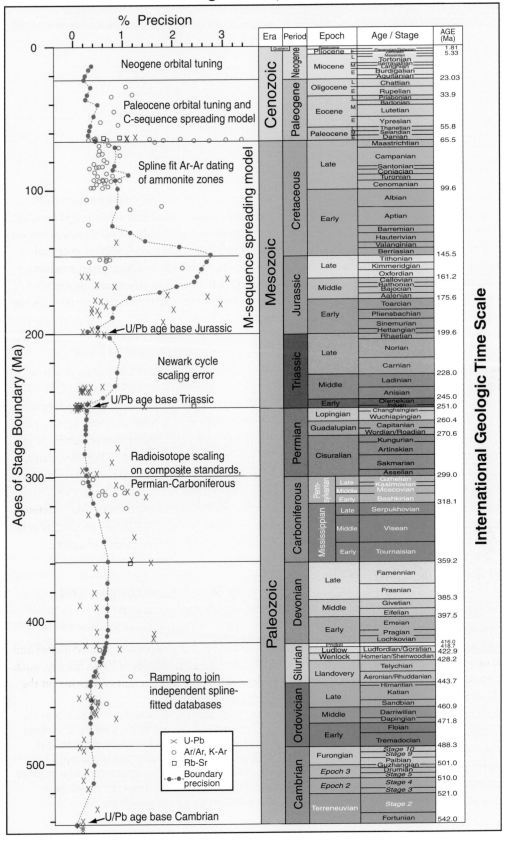

Resolution of Geologic Time (GTS2004 uncertainties)

inter-calibration of events, their relationships to international divisions of geologic time, and their estimated numerical ages.

On-screen display and production of user-tailored time-scale charts is provided by the *TimeScale Creator*, a public JAVA package available from the ICS website (www.stratigraphy. org). In addition to screen views and a scalable-vector graphics (SVG) file for importation into popular graphics programs, the on-screen display has a variety of display options and "hot-curser points" to open windows providing additional information on events, zones, and boundaries.

The database and visualization package are envisioned as a convenient reference tool, chart-production assistant, and a window into the geologic history of our planet. They will be progressively enhanced through the efforts of the subcommissions of the ICS and other stratigraphic and regional experts.

A Geologic Time Scale 2010

At the time of this writing, a major comprehensive update of the Geologic Time Scale is under way, targeted for publication in 2010 in collaboration with Cambridge University Press. All international boundaries (GSSPs) should be established by that date. The book will be an enhanced, improved, and expanded version of GTS04, including chapters on planetary scales, the Cryogenian–Ediacaran periods/systems, a prehistory scale of human development, a survey of sequence stratigraphy, and an extensive compilation of stable-isotope chemostratigraphy.

Age assignments will utilize revised inter-calibration standards and error analysis for different methods of radiogenic isotope analyses. The entire Cenozoic and significant portions of the Mesozoic will have high-resolution scaling based on astronomical tuning or orbital cycles.

Acknowledgements

Individual chapters or diagrams for this book were contributed, extensively revised, or carefully reviewed by subcommission officers of the International Commission on Stratigraphy and other specialists. Some of these contributors are recognized at the end of each chapter, but many other geoscientists provided their expertise. For further details/information on each interval, we recommend the chapters in GTS04. Alan Smith, a co-author on GTS04, provided general advice. Christopher Scotese produced paleogeographic maps for each time slice. Stan Finney, ICS Vice-Chair (and incoming ICS Chair in August, 2008), extensively reviewed the entire draft, especially clarifying the usage of rock/time terminology and status of some pending international stratigraphic units. Susan Francis and Matt Lloyd at Cambridge University Press supervised the production of this book.

Further reading

Dawkins, R., 2004. *The Ancestor's Tale: A Pilgrimage to the Dawn of Life*. London: Weidenfeld & Nicolson.

Gradstein, F.M., Ogg, J.G., Smith, A.G. (coordinators), Agterberg, F.P., Bleeker, W.,

Cooper, R.A., Davydov, V., Gibbard, P., Hinnov, L.A., House, M.R. (†), Lourens, L., Luterbacher, H.-P., McArthur, J., Melchin, M.J., Robb, L.J., Sadler, P.M., Shergold, J., Villeneuve, M., Wardlaw, B.R., Ali, J., Brinkhuis, H., Hilgen, F.J., Hooker, J., Howarth, R.J., Knoll, A.H., Laskar, J., Monechi, S., Powell, J., Plumb, K.A., Raffi, I., Röhl, U., Sanfilippo, A., Schmitz, B., Shackleton, N.J., Shields, G.A., Strauss, H., Van Dam, J., Veizer, J., van Kolfschoten, Th., and Wilson, D., 2004. *A Geologic Time Scale 2004*. Cambridge: Cambridge University Press.

Gradstein, F.M., and Ogg, J.G., 2006. Chronostratigraphic data base and visualization: Cenozoic–Mesozoic–Paleozoic integrated stratigraphy and user-generated time scale graphics and charts. *GeoArabia*, **11**(3): 181–184.

Harland, W.B., Armstrong, R.L., Cox, A.V., Craig, L.E., Smith, A.G., and Smith, D.G., 1989. *A Geologic Time Scale 1989*. Cambridge: Cambridge University Press. [and their previous *A Geologic Time Scale 1982*]

McLaren, D.J., 1978. Dating and correlation: a review. In: *Contributions to the Geologic Time Scale*, Studies in Geology no. 6, eds. G.V. Cohee, M.F. Glaessner, and H. D. Hedberg. Tulsa: American Association of Petroleum Geologists, pp. 1–7.

Remane, J., 2003. Chronostratigraphic correlations: their importance for the definition of geochronologic units. *Palaeogeography, Palaeoclimatology, Palaeoecology*, **196**: 7–18.

Remane, J., Bassett, M.G., Cowie, J.W., Gohrbandt, K.H., Lane, H.R., Michelsen, O., Wang, N., 1996. Revised guidelines for the establishment of global chronostratigraphic standards by the International Commission on Stratigraphy (ICS). *Episodes*, **19**(3): 77–81.

Selby, D., 2007. Direct rhenium-osmium age of the Oxfordian–Kimmeridgian boundary, Staffin Bay, Isle of Skye, UK, and the Late Jurassic time scale. *Norwegian Journal of Geology*, **47**: 291–299.

Van Couvering, J.A., and Ogg, J.G., 2007. The future of the past: geological time in the digital age. *Stratigraphy*, **4**: 253–257.

Selected on-line references

International Commission on Stratigraphy – *www.stratigraphy.org* – for current status of all stage boundaries, time scale diagrams, *TimeScale Creator*, the *International Stratigraphic Guide*, links to subcommission websites, etc.

NOTE: There are many excellent books on historical geology, paleontology, individual periods of geologic time, and other aspects of stratigraphy. Some of this information on the history of Earth's surface and its life is now available on websites which are continuously being updated and enhanced. Some selected ones (biased slightly toward North America) are:

Palaeos: The Trace of Life on Earth (compiled and maintained by Toby White) – *www.palaeos. com* – and other websites that it references at end of each period. There is also a WIKI version being compiled at Palaeos.org. The Palaeos suite has incredible depth and is written for the general scientist.

Smithsonian Institution paleobiology site – *paleobiology.si.edu/geotime/introHTML/index.htm* – After entering, then select Period or Eon by clicking on [*Make a Selection*] in upper right corner of screen.

Web Geological Time Machine (compiled by Museum of Paleontology, University of California) – *www.ucmp.berkeley.edu/exhibits/geologictime.php* – and an accompanying History of Life through Time – *www.ucmp.berkeley.edu/exhibits/historyoflife.php*.

Wikipedia online encyclopedia (a public effort) – *en.wikipedia.org/wiki/Geologic_time_scale* – has excellent reviews of each geologic period and most stages.

Historical Geology on-line (Pamela J.W. Gore, for University System of Georgia) – *gpc.edu/~pgore/geology/historical_lecture/historical_outline.php* – Great image-illustrated site, plus lots of links to other relevant sites from Index page.

Plate Reconstructions (images and animations), some selected sites: Paleomap Project (by Christopher Scotese) – *www.scotese.com/*. Global Plate Tectonics and Paleogeography (Ron Blakey, Northern Arizona University) – *jan.ucc.nau.edu/~rcb7/*, both global and paleogeography of the southwestern USA. Plates (Institute of Geophysics, University of Texas at Austin) – *www.ig.utexas.edu/research/projects/plates/*. Geology: Plate Tectonics (compiled by Museum of Paleontology, University of California) – *www.ucmp.berkeley.edu/geology/tectonics.html*.

EarthTime (maintained by Samuel Bowring, MIT) – *www.earth-time.org/* – information on radiometric dating. CHRONOS (maintained by Cinzia Cervato, Iowa State University) – *www.chronos.org* – databases, especially micropaleontology. Paleobiology Database (maintained by John Alroy) – *paleodb.org/* – mainly macrofossils.

Additional collections of links to stratigraphy of different periods and paleontology of various phyla are at *www.geologylinks.com*, and other sites. The World Wide Web array of posted information grows daily.

Authors

James G. Ogg, Department of Earth and Atmospheric Sciences, Purdue University, 550 Stadium Mall Drive, West Lafayette, IN 47907, USA (Secretary-General, International Commission on Stratigraphy)

Gabi M. Ogg, 1224 N. Salisbury, West Lafayette, IN 47906, USA

Felix M. Gradstein, Geology Museum, University of Oslo, N-0318 Oslo, Norway (Chair, International Commission on Stratigraphy)

2 Planetary time scale

Kenneth L. Tanaka and William K. Hartmann

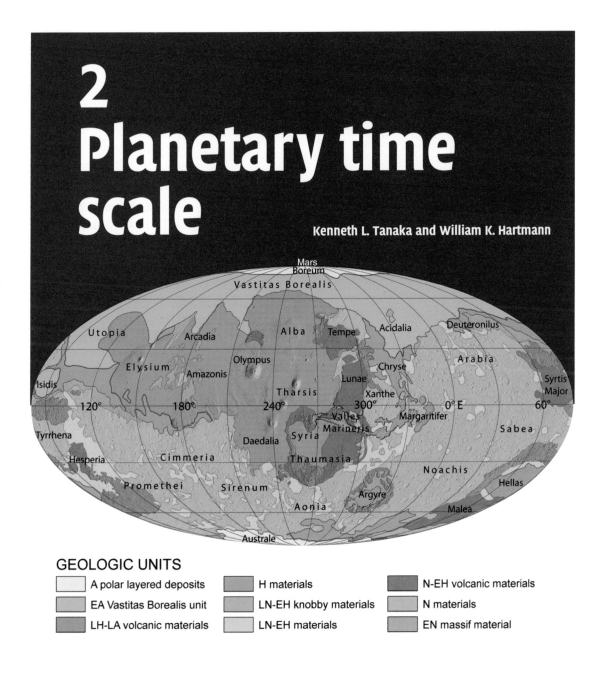

GEOLOGIC UNITS

- A polar layered deposits
- EA Vastitas Borealis unit
- LH-LA volcanic materials
- H materials
- LN-EH knobby materials
- LN-EH materials
- N-EH volcanic materials
- N materials
- EN massif material

Introduction

Figure 2.1. Global geologic map of Mars. Reprinted, with permission, from Nimmo and Tanaka (2005), ©2005 Annual Reviews.

Formal stratigraphic systems have been developed for the surfaces of Earth's Moon, Mars, and Mercury. The time scales are based on regional and global geologic mapping, which establishes relative ages of surfaces delineated by superposition, transaction, morphology, and other relations and features. Referent map units are used to define the commencement of events and periods for definition of chronologic units.

Relative ages of these units in most cases can be confirmed using size–frequency

Figure 2.2. Lunar stratigraphy: Copernicus region of the Moon. Approximate location of this region is shown on a photograph of the Moon provided by Gregory Terrance (Finger Lakes Instrumentation, Lima, New York; www.fli-cam. com). Copernicus crater (C) is 93 km in diameter and centered at 9.7°N, 20.1°W. Copernicus is representative of bright-rayed crater material formed during the lunar Copernican Period. Its ejecta and secondary craters overlie Eratosthenes crater (E), which is characteristic of relatively dark crater material of the Eratosthenian Period. In turn, Eratosthenes crater overlies relatively smooth mare materials (M) of the Late Imbrian Epoch. The oldest geologic unit in the scene is the rugged rim ejecta of Imbrium basin (I), which defines the base of the Early Imbrian Epoch. (Lunar Orbiter IV image mosaic; north at top; illumination from right; courtesy of U.S. Geological Survey Astrogeology Team.)

distributions and superposed craters. For the Moon, the chronologic units and cratering record are constrained by radiometric ages measured from samples collected from the lunar surface. This allows a calibration of the areal density of craters vs. age, which permits model ages to be measured from crater data for other lunar surface units. Model ages for other cratered planetary surfaces are constructed by two methods: (1) estimating relative cratering rates with Earth's Moon and (2) estimating cratering rates directly based on surveys of the sizes and trajectories of asteroids and comets.

The Moon

The first formal extraterrestrial stratigraphic system and chronology was developed for Earth's Moon beginning in the 1960s, first based on geologic mapping using telescopic observations. These early observations showed that the rugged lunar highlands are densely

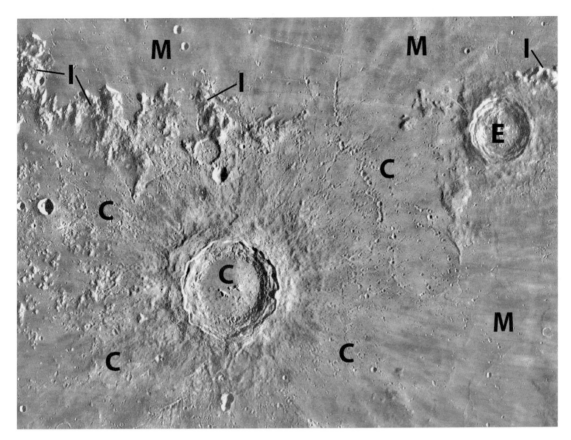

Figure 2.2. (cont.)

cratered, whereas the *maria* (Latin for "seas") form relatively dark, smooth plains consisting of younger deposits that cover the floors of impact basins and intercrater plains. Resolving power of the lunar landscape improved greatly with the Lunar Orbiter spacecraft, which permitted also the first mapping of the farside of the Moon. By the end of the decade and into the 1970s, manned and unmanned exploration of lunar sites by the Apollo and Luna missions brought return of samples. The majority of early exploration involved the lunar nearside (facing Earth), and the stratigraphic system and chronology follow geologic features and events primarily expressed on the nearside. Based on geologic inferences, returned samples were used to date with radiometric methods the materials of the early crust and the emplacement of extensive lava flows that make up the lunar maria. Attempts were also made to use the samples to date certain lunar basin-forming impacts and the large craters Copernicus and Tycho. Two processes have mainly accomplished resurfacing: impacts and volcanism. Analogous to volcanism, impact heating can generate flow-like deposits of melted debris that can infill crater floors or terrains near crater rims. As on Earth, the broadest time intervals are designated "Periods" and their subdivisions are "Epochs" (if not meeting formal stratigraphic criteria, these unit categories are not capitalized).

From oldest to youngest, lunar chronologic units and their referent surface materials and events include:

(1) *pre-Nectarian* period, earliest materials dating from solidification of the crust (a suite of anorthosite, norite, and troctolite) until just before formation of Nectaris basin;

(2) *Nectarian* Period, mainly impact melt and ejecta associated with Nectaris basin and later impact features;

(3) *Early Imbrian* Epoch, consisting mostly of basin-related materials associated at the beginning with Imbrium basin and ending with Orientale basin;

(4) *Late Imbrian* Epoch, characterized by mare basalts post-dating Orientale basin;

(5) *Eratosthenian* Period, represented by dark, modified ejecta of Eratosthenes crater; and

(6) *Copernican* Period, characterized by relatively fresh bright-rayed ejecta of Copernicus crater.

The cratering rate was initially very high; uncertain is whether or not the lunar cratering rate records a relatively brief period of catastrophic bombardment in the inner solar system at ~4.0 Ga, possibly spawned by perturbations in the orbits of the giant outer planets. Alternatively, the dense population of highland craters records the gradual trailing off of the accretional period itself. Telescopic surveys of the numbers, sizes, and orbits of asteroids indicate that they have been the prime contributor to the lunar cratering record.

Mars

The Red Planet has a geologic character similar to the Moon, with vast expanses of cratered terrain and lava plains, but with the important addition of features resulting from the activity of wind and water over time. This results in a geologically complex surface history; geologic mapping has assisted in unraveling it, following the approaches developed for studies of the Moon. Beginning in the 1970s with the Mariner 9 and Viking spacecraft, and continuing with a flotilla of additional orbiters and landers beginning in the 1990s, Mars has become a highly investigated planet. Geologic mapping led to characterization of periods and epochs as on the Moon.

The *pre-Noachian* period represents the age of the early crust and is not represented in known outcrops, but a Martian meteorite, ALH84001, was crystallized at ~4.5 Ga.

Heavily cratered terrains formed during the *Noachian* Period. These include large impact

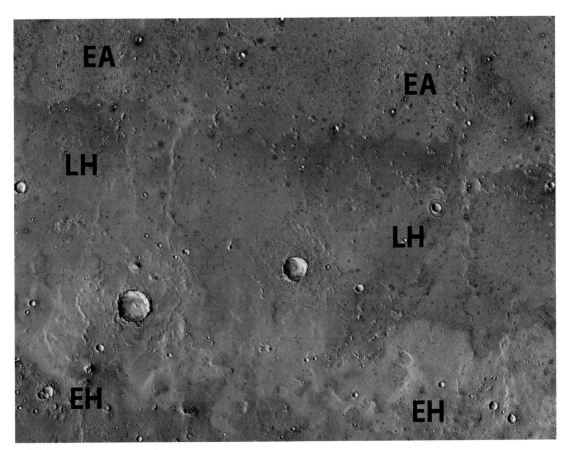

Figure 2.3. Martian stratigraphy: part of south-central Utopia Planitia in the northern lowlands of Mars. Image base consists of (1) a partly transparent Thermal Emission Imaging System (THEMIS) daytime infrared image mosaic (~230 m/pixel) in which brightness indicates surface temperature, overlying (2) a color shaded-relief digital elevation model from Mars Orbiter Laser Altimeter (MOLA) data (brown is high, purple is low; ~460 m/pixel). Relatively bright (i.e., warmer), finely ridged, and hummocky Early Amazonian plains-forming material (EA) defines the base of the Amazonian Period on Mars. This material overlies smooth, locally knobby and ridged Late Hesperian material (LH) that in turn embays depressions and scarps marking the rolling and hollowed surface of yet older, Early Hesperian material (EH). (View centered near 19°N, 113°E; 412 km scene width; north at top; illumination from lower left; THEMIS global mosaic courtesy of Christensen, P.R., N.S. Gorelick, G.L. Mehall, and K.C. Murray, *THEMIS Public Data Releases*, Planetary Data System node, Arizona State University, http://themis-data.asu.edu; MOLA data courtesy of MOLA Science Team.)

basins of the Early Noachian Epoch, vast cratered plains of the Middle Noachian, and intercrater plains resurfaced by fluvial and possibly volcanic deposition during the Late Noachian when the atmosphere apparently was thicker and perhaps warmer and heat flow was higher.

Hesperian Period rocks are much less cratered and record waning fluvial activity but extensive volcanism, particularly during the Early Hesperian Epoch. Mars Express and Mars Reconnaissance Orbiter data indicate that clay minerals occur in some Noachian strata, whereas hydrated sulfates are mostly in

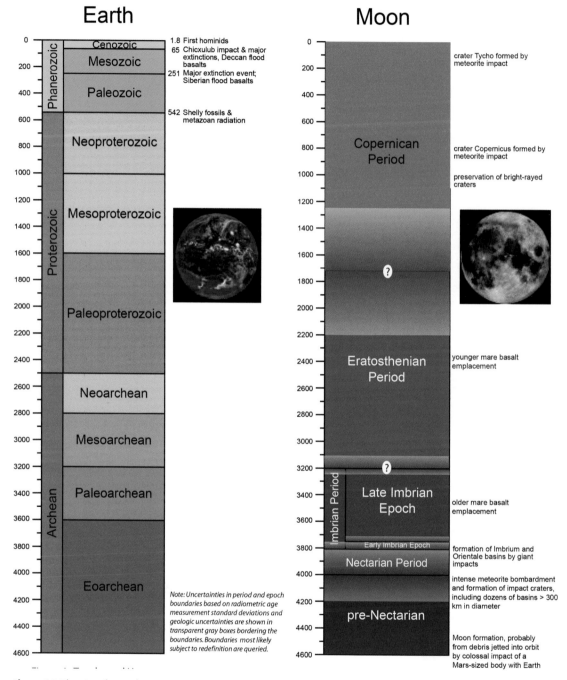

Earth

0		Cenozoic	1.8 First hominids
			65 Chicxulub impact & major extinctions, Deccan flood basalts
200		Mesozoic	
400	Phanerozoic	Paleozoic	251 Major extinction event; Siberian flood basalts
600			542 Shelly fossils & metazoan radiation
800		Neoproterozoic	
1000			
1200			
1400	Proterozoic	Mesoproterozoic	
1600			
1800			
2000		Paleoproterozoic	
2200			
2400			
2600		Neoarchean	
2800			
3000		Mesoarchean	
3200			
3400	Archean	Paleoarchean	
3600			
3800			
4000		Eoarchean	
4200			
4400			
4600			

Note: Uncertainties in period and epoch boundaries based on radiometric age measurement standard deviations and geologic uncertainties are shown in transparent gray boxes bordering the boundaries. Boundaries most likely subject to redefinition are queried.

Moon

0		
200		crater Tycho formed by meteorite impact
400		
600		
800	Copernican Period	crater Copernicus formed by meteorite impact
1000		preservation of bright-rayed craters
1200		
1400		
1600	?	
1800		
2000		
2200		
2400	Eratosthenian Period	younger mare basalt emplacement
2600		
2800		
3000		
3200	?	
3400	Imbrian Period — Late Imbrian Epoch	older mare basalt emplacement
3600		
3800	Early Imbrian Epoch	formation of Imbrium and Orientale basins by giant impacts
4000	Nectarian Period	intense meteorite bombardment and formation of impact craters, including dozens of basins > 300 km in diameter
4200		
4400	pre-Nectarian	Moon formation, probably from debris jetted into orbit by colossal impact of a Mars-sized body with Earth
4600		

Figure 2.4. Planetary time scales.

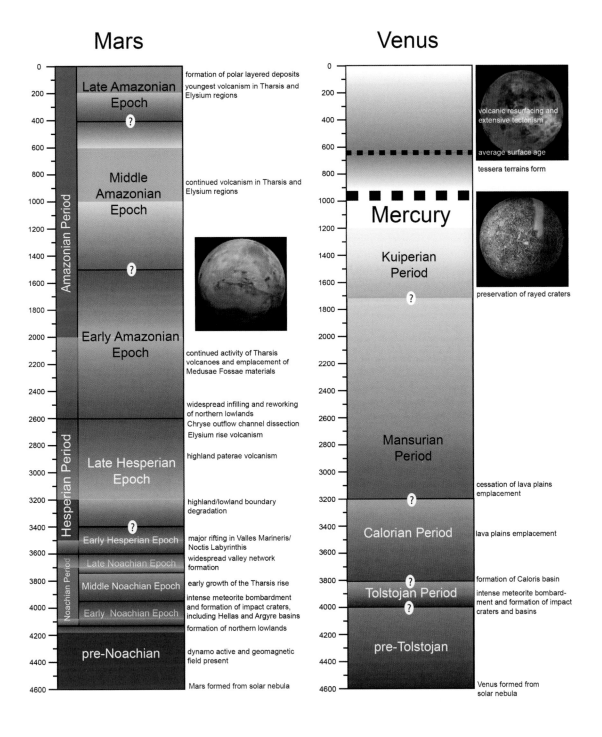

Mars

Late Amazonian Epoch		formation of polar layered deposits youngest volcanism in Tharsis and Elysium regions
Middle Amazonian Epoch	Amazonian Period	continued volcanism in Tharsis and Elysium regions
Early Amazonian Epoch		continued activity of Tharsis volcanoes and emplacement of Medusae Fossae materials
Late Hesperian Epoch	Hesperian Period	widespread infilling and reworking of northern lowlands Chryse outflow channel dissection Elysium rise volcanism highland paterae volcanism highland/lowland boundary degradation
Early Hesperian Epoch		major rifting in Valles Marineris/ Noctis Labyrinthis
Late Noachian Epoch	Noachian Period	widespread valley network formation
Middle Noachian Epoch		early growth of the Tharsis rise
Early Noachian Epoch		intense meteorite bombardment and formation of impact craters, including Hellas and Argyre basins formation of northern lowlands
pre-Noachian		dynamo active and geomagnetic field present Mars formed from solar nebula

Venus

volcanic resurfacing and extensive tectonism

average surface age

tessera terrains form

Mercury

Kuiperian Period	preservation of rayed craters
Mansurian Period	cessation of lava plains emplacement
Calorian Period	lava plains emplacement
Tolstojan Period	formation of Caloris basin intense meteorite bombardment and formation of impact craters and basins
pre-Tolstojan	Venus formed from solar nebula

Hesperian rocks. A thick permafrost zone developed as the surface cooled, and much of the fluvial activity during the Late Hesperian Epoch occurred as catastrophic flood outbursts through this frozen zone, perhaps initiated by magmatic activity.

The *Amazonian* Period began with expansive resurfacing of the northern lowlands, perhaps by sedimentation within a large body of water. Much lower levels of volcanism and fluvial discharges, coupled with aeolian deposition and erosion continued into the Middle and Late Amazonian Epochs. Continued weathering has led to iron oxidation of surface materials.

The polar plateaus, covered by bright deposits of residual ice as well as seasonally waxing and waning meter-thick CO_2 frost, are among the youngest features on the planet. Ice-rich mantles and glacial-like deposits at middle and equatorial latitudes signal climate oscillations in the relatively recent geologic record.

Mercury

The innermost planet was partly imaged by flybys of the Mariner 10 spacecraft in 1974 and 1975, enabling stratigraphic studies that reveal a remarkably similar surface history to that of Earth's Moon. Consequently, a Mercurian chronology was developed based on impact basins and craters that may have similar histories to comparable lunar features.

Thus, five major periods have been proposed that correspond to those of the Moon, as follows:

(1) *pre-Tolstojan* (equivalent to the lunar pre-Nectarian)

(2) *Tolstojan* (Nectarian)

(3) *Calorian* (Imbrian)

(4) *Mansurian* (Eratosthenian)

(5) *Kuiperian* (Copernican).

Absolute ages for these periods are much more uncertain than for the Moon and Mars.

Venus

The Venusian surface has been investigated extensively with orbiters and landers, most recently by the Magellan orbiter with its mapping radar in the 1990s. Impact crater densities are low. Statistics of nearly a thousand impact craters on its surface indicate that Venus has an average surface age of hundreds of millions of years. In spite of its spectacular volcanic surface dotted with thousands of volcanoes and broad fields of lava flows, all of which has been tectonically disrupted to varying degrees, the details of the global geologic evolution of this Earth's twin planet in size are not well constrained. Possibilities range from local to regional events driven by mantle plumes to global volcanic and tectonic evolution driven by atmospheric greenhouse-heating effects on Venusian climate.

Other Solar System bodies

The solid surfaces of asteroids and satellites of Jupiter, Saturn, Uranus, and Neptune show varying degrees of cratering that reflect surface ages. While asteroids are commonly saturated with craters, indicating their primordial origin, some asteroids, comet nuclei, and other bodies demonstrate later resurfacing as their rocky or icy crusts evolved. Dating these surfaces relies on inferences of the populations of projectiles across time and space. Absolute dates are very poorly constrained. Complications in estimates of cratering rates include the relative importance of asteroids in the inner solar system versus that of comets and other icy materials of the Kuiper Belt.

Further reading

Basaltic Volcanism Study Project, 1981. *Basaltic Volcanism on the Terrestrial Planets*. Houston: Lunar and Planetary Institute.

Bougher, S.W., Hunten, D.M., and Phillips, R.J. (eds.), 1997. *Venus II: Geology, Geophysics, Atmosphere, and Solar Wind Environment*. Tucson: University of Arizona Press.

Hartmann, W.K., 2005. Martian cratering 8: isochron refinement and the chronology of Mars. *Icarus*, **174**: 294–320.

Kallenbach, R., Geiss, J., and Hartmann, W.K. (eds.), 2001. *Chronology and Evolution of Mars*. Dordrecht: Kluwer Academic Publishers.

Nimmo, F., and Tanaka, K., 2005. Earth crustal evolution of Mars. *Annual Review of Earth and Planetary Sciences*, **33**: 133–161.

Schenk, P.M., Chapman, C.R., Zahnle, K., and Moore, J.M., 2004. Ages and interiors: the cratering record of the Galilean satellites. In: *Jupiter: The Planet, Satellites and Magnetosphere*, eds. F. Bagenal, T.E. Dowling, and W.B. McKinnon. Cambridge: Cambridge University Press, pp. 427–456.

Shoemaker, E.M., and Hackman, R.J., 1962. Stratigraphic basis for a lunar time scale. In: *The Moon*, eds. Z. Kopal and Z. K Mikhailov. London: Academic Press, pp. 289–300.

Spudis, P.D., and Guest, J.E., 1988. Stratigraphy and geologic history of Mercury. In: *Mercury*, eds. F. Vilas, C.R. Chapman, and M.S. Matthews. Tucson: University of Arizona Press, pp. 118–164.

Stöffler D., and Ryder, G., 2001. Stratigraphy and isotope ages of lunar geologic units: chronological standards for the inner solar system. *Space Science Reviews*, **96**: 9–54.

Strom, R.G., Malhotra, R., Ito, T., Yoshida, F., and Kring, D.A., 2005. The origin of planetary impactors in the inner solar system. *Science*, **309**: 1847–1850.

Tanaka, K.L., 1986. The stratigraphy of Mars. Proceedings of the Lunar and Planetary Science Conference, 17, part 1,

Journal of Geophysical Research, **91**: E139–E158.

Wilhelms, D.E., 1987. *The Geologic History of the Moon*, U.S. Geological Survey Professional Paper, no. 1348. Washington, DC: U.S. Government Printing Office.

Selected on-line references and imagery

U.S. Geological Survey Astrogeology Research Program – *astrogeology.usgs.gov/*, especially: Browse the Solar System: *astrogeology.usgs.gov/Projects/BrowseTheSolarSystem/*

Solar System Exploration (NASA) – *solarsystem.nasa.gov/index.cfm*

Welcome to the Planets (JPL, NASA) – *pds.jpl.nasa.gov/planets/*

Mars Exploration Program (NASA) – *marsprogram.jpl.nasa.gov/*

Wikipedia – Lunar geologic timescale – *en.wikipedia.org/wiki/Lunar_geologic_time_scale*

Gregory Terrance (Finger Lakes Instrumentation, Lima, New York; *www.fli-cam.com*) Planetary Data System node, Arizona State University, *http://themis-data.asu.edu*

Authors

William K. Hartmann, Planetary Science Institute, 1700 East Fort Lowell, Suite 106, Tucson, AZ 85719, USA

Kenneth L. Tanaka, Astrogeology Team, U.S. Geological Survey, 2255 North Gemini Drive, Flagstaff, AZ 86001, USA

3 Precambrian

Martin J. Van Kranendonk,
with the Cryogenian-Ediacaran portions by
James Gehling and Graham Shields

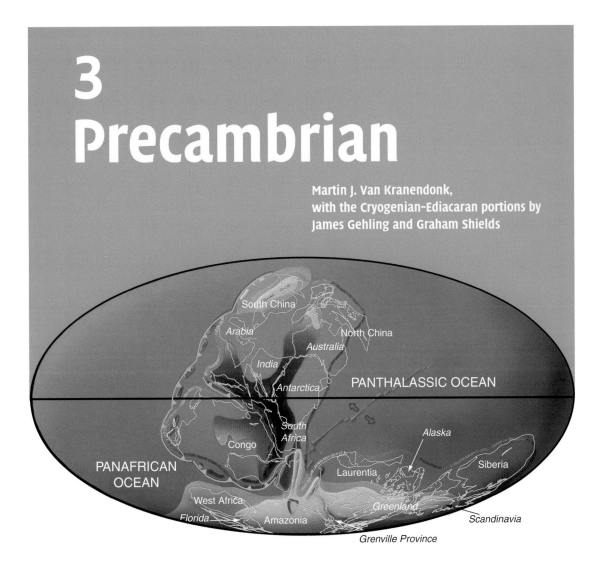

Figure 3.1. Geographic distribution of the continents at approximately the end of the Cryogenian Period (650 Ma). The paleogeographic map was provided by Christopher Scotese, although the positions of continents are uncertain.

History and subdivisions of Precambrian

The *Precambrian* is not a formal stratigraphic unit, but simply refers to all rocks that formed prior to the beginning of the Cambrian Period (base of Phanerozoic) and, by its very nature, back to the formation of Earth.

The lack of a diverse and well-preserved fossil record, the generally decreasing volume of supracrustal rocks, and increasing degree of metamorphism and tectonic disturbance, as well as the uncertainties in the configuration of the continents, all contribute to the challenge of establishing a chronostratigraphic time scale beyond the Phanerozoic Eon. Therefore, the main method for correlation of Precambrian strata requires radiometric ages of interbedded volcanic rocks and plutonic rocks. Accordingly, in 1990, the IUGS ratified the recommendations by the Precambrian Subcommission that the subdivisions of the Archean and Proterozoic eons shall be defined chronometrically, with boundaries assigned in terms of a round number of millions of years before present (Ma) or *Global Standard Stratigraphic Ages* (GSSAs). This scheme was viewed as the most appropriate solution until packages of strata and associated global events could be recognized and correlated by the intrinsic features of their geologic history rather than simply by numerical dates.

The best age for the formation of the Earth comes from considering meteorites as part of the same evolving system, and these indicate an approximate age of accretion for the Solar System of 4.567 Ga. The Precambrian rock record is extremely sparse prior to 3.8 Ga, with only a handful of ancient cratons containing any evidence for older crust. As a result, the base of the Archean Eon is roughly indicated at 4.0 Ga. The Archean Eon is subdivided into four eras (Fig. 3.2, left side).

The Proterozoic Eon begins at 2.5 Ga, which is the approximate time when most granite–greenstone crust had formed, when oxygen from cyanobacteria began to dramatically change Earth's atmosphere and ocean chemistry, and when complex one-celled life (eukaryotes) evolved from simple cells (prokaryotes). The Proterozoic is subdivided into 10 periods, generally of 200-myr durations, grouped into three eras (Fig. 3.2, left side). These Proterozoic periods and their nomenclature were chosen to reflect large-scale tectonic or sedimentary features that occurred within each period (Table 3.1). For example, the Siderian Period (2.5–2.3 Ga) is named from the banded-iron deposits (*sideros* = iron), which peaked within that interval and were accompanied by the near-global Huronian glaciation.

Several features of the current ICS stratigraphic chart relating to the Precambrian time scale have raised concern within the geological community, primary among which is that the boundaries are based purely on round-number chronometric divisions and ignore stratigraphy. Since 1990, there has been a veritable explosion of new geoscience information on Precambrian terranes and geobiology, including many thousands of precise U–Pb zircon age dates. Another powerful tool for precise correlation of pre-fossiliferous strata is the emerging recognition of pronounced excursions in carbon isotopes, coupled with trends in sulfur and strontium isotope ratios. Cyanobacteria, especially as expressed by their stromatolite constructions, are widespread in late Archean and Proterozoic strata, and show broad patterns of change in form and microfabric. Acritarchs, spherical to polygonal, organic-walled microfossils, may enable a broad zonation of the Ediacaran and older Neoproterozoic strata.

The Precambrian subcommission is striving to establish a more "natural" set of subdivisions that incorporates major tectonic, biologic, atmospheric, and geochemical events. It is felt that, where possible, type sections and GSSPs should be applied to major Precambrian time scale boundaries. This aim was partly accomplished in 2004, when the IUGS ratified a boundary stratotype (GSSP) for the base of the Ediacaran Period, the youngest period/system of the Proterozoic Eon. The Neoproterozoic subcommission is currently engaged in assessing

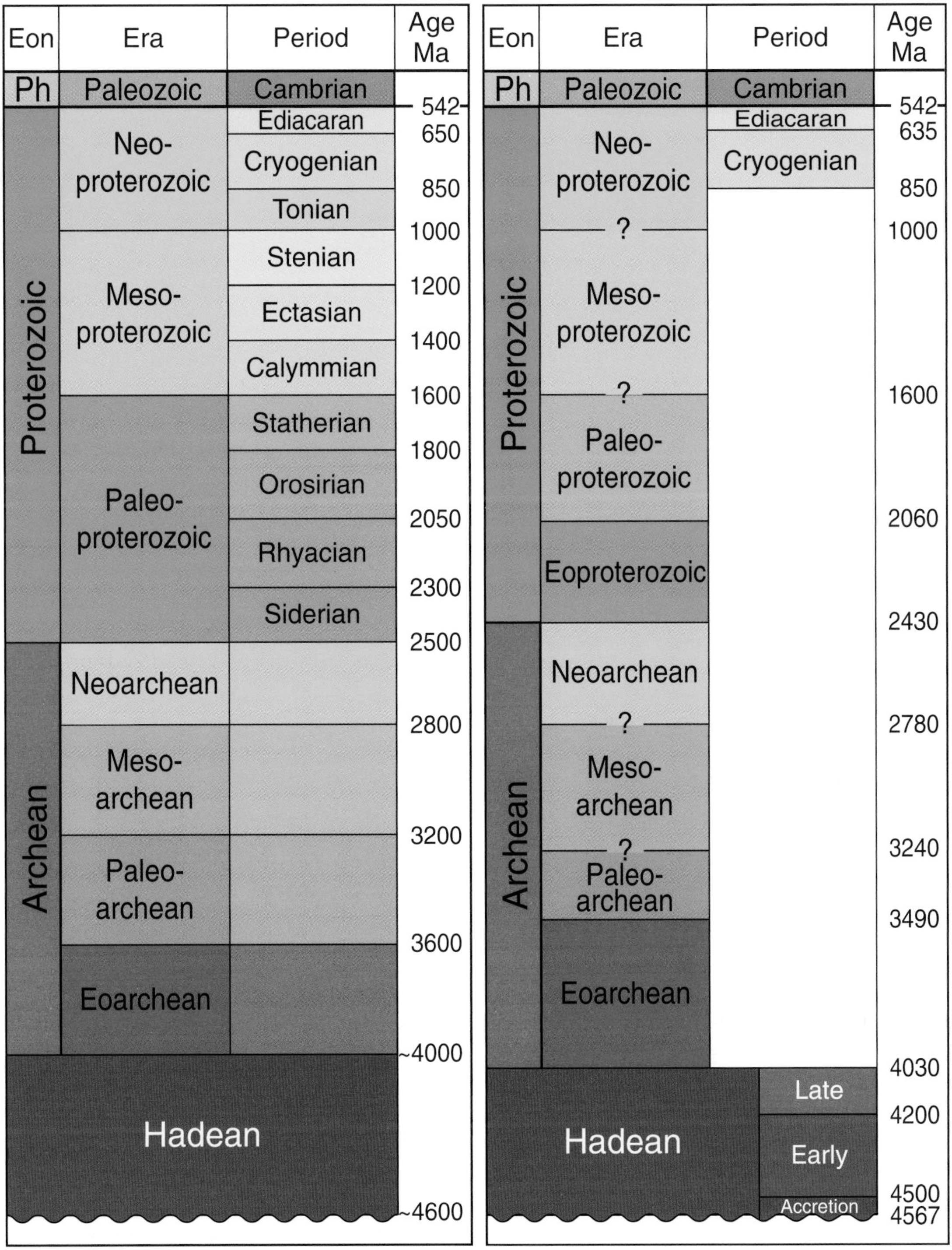

Eon	Era	Period	Age Ma	Eon	Era	Period	Age Ma
Ph	Paleozoic	Cambrian	542-	Ph	Paleozoic	Cambrian	542-
	Neo-proterozoic	Ediacaran	650		Neo-proterozoic	Ediacaran	635
		Cryogenian	850			Cryogenian	850
		Tonian	1000		?		1000
	Meso-proterozoic	Stenian	1200		Meso-proterozoic		
		Ectasian	1400				
		Calymmian	1600		?		1600
	Paleo-proterozoic	Statherian	1800		Paleo-proterozoic		
		Orosirian	2050				2060
		Rhyacian	2300		Eoproterozoic		
		Siderian	2500				2430
	Neoarchean		2800		Neoarchean		2780
	Meso-archean				?		
			3200		Meso-archean		3240
	Paleo-archean				?		
			3600		Paleo-archean		3490
	Eoarchean		~4000		Eoarchean		4030
	Hadean		~4600		Hadean	Late	4200
						Early	4500
						Accretion	4567

Figure 3.2. Current International Stratigraphic Chart for the Precambrian (left) and some of the changes to the Precambrian time scale under consideration, as summarized in this chapter (right).

Table 3.1 Explanation of nomenclature used at the period level in the Proterozoic Eon

Period name	Derivation and geological process	
Ediacaran	Australian Aborigine term referring to a place where water is or was present close by	*Earliest metazoan life*
Cryogenian	*Cryos* = ice; *genesis* = birth Glacial deposits, which typify the late Proterozoic, are most abundant during this interval	*Global glaciation*
Tonian	*Tonas* = stretch Further major platform cover expansion (e.g., Upper Riphean, Russia.; Qingbaikou, China; basins of northwest Africa), following final cratonization of polymetamorphic mobile belts, below	
Stenian	*Stenos* = narrow Narrow polymetamorphic belts, characteristic of the mid-Proterozoic, separated the abundant platforms and were orogenically active at about this time (e.g., Grenville, Central Australia)	*Narrow belts of intense metamorphism and deformation*
Ectasian	*Ectasis* = extension Platforms continue to be prominent components of most shields	*Continued expansion of platform covers*
Calymmian	*Calymma* = cover Characterized by expansion of existing platform covers, or by new platforms on recently cratonized basement (e.g., Riphean of Russia)	*Platform covers*
Statherian	*Statheros* = stable, firm This period is characterized on most continents by either new platforms (e.g., North China, North Australia) or final cratonization of fold belts (e.g., Baltic Shield, North America)	*Stabilization of cratons; cratonization*
Orosirian	*Orosira* = mountain range The interval between about 1900 Ma and 1850 Ma was an episode of orogeny on virtually all continents	*Global orogenic period*
Rhyacian	*Rhyax* = stream of lava The Bushveld Complex (and similar layered intrusions) is an outstanding event of this time	*Injection of layered complexes*
Siderian	*Sideros* = iron The earliest Proterozoic is widely recognized for an abundance of BIF, which peaked just after the Archean-Proterozoic boundary	*Banded-iron formations* (BIF)

Source: Modified from Table 9.1 in *A Geologic Time Scale 2004*.

the possible criteria for establishing a base to the Cryogenian Period.

Beginning of the Archean (4.0 Ga)

A stratigraphic chart – as its name implies – should be based on an existing rock record. The oldest dated rock on Earth is 4.03 Ga from the Acasta gneiss complex of the Slave Craton, northwestern Canada (Stern and Bleeker, 1998). Until further discoveries push back this age, this could be considered as the beginning of the Archean Eon, the oldest major *chronostratigraphic* division of the time scale, and is a chronometric time marker, or GSSA (Fig. 3.2, right side).

The Hadean (4.6 to 4.0 Ga): the hidden earliest 600 myr of Earth's history

Prior to the age of the oldest preserved rock, the Earth had already undergone a complex history spanning over ~537 myr, the equivalent duration of the whole of the Phanerozoic. Many refer to this interval as the *Hadean*, following Cloud (1972). It is felt that some kind of formal definition is warranted, but without placing it within the context of Eon or Era.

It is now well documented that the condensation of solid material to form the terrestrial planets in our Solar System occurred at 4.567 Ga (or T_0), and that accretionary processes continued for ~30–100 myr thereafter, including the formation of the Moon as a result of the glancing impact of a Mars-size planet called *Theia*, about 40 myr after T_0.

At 4.50 Ga, Earth was a molten ball, but a dominantly basaltic crust quickly covered it. Ancient detrital zircons found in Archean rocks from the Jack Hills, Western Australia, provide evidence for water on Earth and for the formation of at least some felsic crust as early as 4.4 Ga (Wilde *et al.*, 2001). Xenocrystic and detrital zircons in Archean rocks recycled from pieces of more ancient crust, together with isotopic data, indicate that crust continued to form throughout the period from 4.4 to 4.03 Ga, but that none of it survived the conditions on young Earth.

Potential subdivisions of the Hadean might include (Fig. 3.2, right side):

- A stage of accretion and differentiation, from T_0 at 4.567 Ga to 4.50 Ga.

- An Early and a Late Hadean, based on evidence from oxygen isotopes in the most ancient (4.4–4.0 Ga) zircons that show a dramatic shift in composition at 4.2 Ga (Cavosie *et al.*, 2005). This may reflect a change from hot conditions prior to 4.2 Ga to cooler conditions thereafter.

Eoarchean Era (4.0 to 3.5 Ga)

The most ancient pieces of Archean crust are highly deformed gneisses dominated by sodic metaplutonic granites (tonalite–trondhjemite–granodiorite) with slivers of highly dismembered mafic and ultramafic rocks and rare metasedimentary rocks. Several remnants of these high-grade gneiss terranes from the period 4.03–3.5 Ga are in cratons around the world, including the Napier Complex in Antarctica, the North China Craton, the Yilgarn Craton (Australia), the Slave Craton (Canada), the Ancient Gneiss Complex in southern Africa, and in West Greenland and Labrador in the dismembered North Atlantic Craton.

This period of crust formation overlaps with an episode of intense meteor bombardment of the Moon, determined from K–Ar dating of volcanic glass beads and Mare basalts collected during Apollo missions as between 4.0–3.85 Ga (Ryder, 2000). Theoretical considerations indicate that the Earth's early crust would have been almost completely destroyed by this *Late Heavy Meteor Bombardment*, and the sparse geological record prior to 3.85 Ga on Earth, together with isotopic data, certainly seems to support this contention (e.g. Kamber, 2007). The postulated larger impacts would have boiled the entire ocean, thereby sterilizing the Earth's surface of evolving life forms, except for types of Archea that inhabited deeper regions of the crust (e.g., Zahnle *et al.*, 2007). Earth's phylogenetic "tree of life" suggests a bottleneck with preferential survival of thermophile Archea that led to the later diversity of life. However, there is no direct evidence for this Late Heavy Meteor Bombardment on Earth, and there is some caution in regard to how to interpret the lunar data.

The oldest preserved supracrustal rocks (sedimentary or volcanic rocks that were deposited on the surface of the Earth) are metamorphosed tectonic slices of ~3.83 Ga and ~3.71 Ga basalts, felsic tuffs, and banded-iron formation of the Isua supracrustal belt and associated Akilia association in west Greenland.

Supracrustal rocks that are well preserved at low metamorphic grade are preserved at about 3.53 Ga in both the Barberton greenstone belt of the Kaapvaal Craton, southern Africa, and in the East Pilbara Terrane of the Pilbara Craton, northwestern Australia. These rocks form the oldest part of both greenstone successions and originally may have been part of one larger protocontinental nucleus.

However, it is not until 3.49 Ga that very well-preserved, demonstrably autochthonous, and more widespread successions are preserved. The best of these, in terms of continuity of stratigraphy, preservation and biological importance, occurs in the North Pole Dome area of the Pilbara Craton. Here, a 12-km thick, continuous succession of low-grade, dominantly volcanic rocks contains an 8–60-m thick sedimentary unit known as the Dresser Formation near the base of the succession. The Dresser Formation hosts evidence of the oldest life on Earth in the form of fossil stromatolites and highly fractionated $\delta^{13}C$ values of kerogen indicative of methane-consuming life. This sedimentary unit of 3.49–3.48 Ga is conformably bound by well-preserved pillow basalts of the Warrawoona Group.

A potential chronostratigraphic boundary marker to mark the end of the *Eoarchean* and the start of the *Paleoarchean* is the base of the stromatolitic Dresser Formation in the North Pole Dome. This site has the advantages of being located within a continuous, well-dated stratigraphic succession with easy access that is part of a well-documented, and well-understood, terrane.

Archean–Proterozoic transition (2.5 Ga)

The Archean–Proterozoic boundary at 2.5 Ga is widely regarded as both useful and significant. It approximates the end of the last major period of granite–greenstone development on Earth, a change in the composition of subcontinental mantle lithosphere, as well as the time of transition to an oxygenated atmosphere.

Geodynamically, the change from a "mobilist Archean regime" to a more stable and more "modern" Proterozoic Earth was a transition that lasted several hundred million years. For example, cratonization of some pieces of crust had occurred by 2.83 Ga (Pilbara Craton), whereas others continued to form to ~2.50 Ga (Dharwar Craton). Indeed, typically "Archean" granite–greenstone crust continued to form to at least 1.9 Ga (e.g., in the Man Craton, West Africa, and in the Flin Flon greenstone belt, Canada). Several key lithological units are also distributed broadly in time across the Archean–Proterozoic boundary. For example, banded-iron formations (BIF) are common and voluminous as early as 2.8 Ga in the Yilgarn Craton, but continue in the rock record to 1.8 Ga; and komatiites are common in 3.5–2.7-Ga granite–greenstone terranes but are also present in 2.056 Ga rocks in Finnish Lapland.

In geobiological terms, many geologists believe that the most significant change in Earth history was the development of an oxygenated atmosphere, as this allowed for the evolution of complex life on Earth, eventually including our species. Oxygenation is widely considered to be the result of the respiration of cyanobacteria, who utilize CO_2 and sunlight to produce food energy (carbon compounds) and give off free oxygen as a waste product. Many well-known geological changes accompanied this change in atmospheric composition, including the disappearance of detrital uraninite and pyrite in sandstones, and the appearance of redbeds and Mn-rich sedimentary rocks (e.g., Melezhik et al., 2005). Many redox-sensitive chemical tracers of this rise have been identified, including Mo isotopes, Ce, Fe, Re–Os data, sulfur isotopes, and platinum group element concentrations (Siebert et al., 2005). Because these proxies have different sensitivities to redox conditions, then ages can be assigned to different stages in the rise of oxygen.

The Archean–Proterozoic boundary only roughly approximates the rise of cyanobacteria and oxygenation of the atmosphere. There is strong evidence that cyanobacteria arose significantly earlier (~2.7 Ga: Brocks et al., 2003), as did the onset of oxygenation of the atmosphere. A critical step in the oxygenation of the atmosphere occurred at ~2.32 Ga with the disappearance of the mass-independently fractionated sulfur isotope signature (S-MIF), which indicates the development of an ozone layer requiring a partial pressure of oxygen of >10^{-5} present atmospheric levels (Farquhar et al., 2000; Papineau et al., 2007). This signal disappears within the period of global, low-latitude Paleoproterozoic glaciations and near the onset of the Lomagundi–Jatuli positive δ^{13}C isotopic excursion (Melezhik et al., 2005).

The predicted drop in partial pressure of CO_2 in the atmosphere across the Archean–Proterozoic boundary should produce a change from predominantly chemical weathering to dominantly physico-mechanical weathering. This change in weathering style is due to a decrease in the concentration of carbonic acid (H_2CO_3), which forms through the combination of H_2O and CO_2 under higher pCO_2 and acts as a chemical weathering agent by increasing the acidity (lower pH) of the world's oceans, rivers and rain.

Two sets of observations from the geological record may show this change. The first set is the horizon where basin deposition shifts from predominantly chemical precipitates (banded-iron formation, dolomite and an unusual type of Fe-rich shale) to clastic sedimentary rocks (fine-grained quartzo-feldspathic turbidites, including shale, siltstone and sandstone, and granular banded-iron formation). In Western Australia, this transition is a sharp conformable contact between the Boolgeeda Iron Formation (2445 ± 5 Ma) at the top of the Hamersley Group and the overlying Turee Creek Group. In the Transvaal

Supergroup of southern Africa, the age for this transition is constrained to between 2465 and 2432 Ma.

The second observation comes from the >10-km thick, rift-related Huronian Supergroup in North America. Three glacial–interglacial cycles are preserved in the lower part of this succession, deposited above 2496–2450 Ma layered gabbro–anorthosite intrusions and volcanics, and cut by the ~2.22 Ga Nipissing diabase. The chemical composition of sandstones indicates a decrease in chemical weathering effects above the second glaciation and is closely coincident with the disappearance of the S-MIF isotopic signal (Fedo *et al.*, 1997; Papineau *et al.*, 2007).

Therefore, it is being considered that, although a "nice round number" chronometric boundary of 2.5 Ga is broadly useful, it does not represent any specific, rock-based event, nor is it particularly scientific. Instead, a chemo- and or litho-stratigraphic boundary could mark the Archean–Proterozoic boundary, at either:

(1) Top of banded-iron formations in the Hamersley Group, Pilbara Craton.

(2) Disappearance of the S-MIF isotopic signal in sedimentary rocks of the Huronian Supergroup, although there may be difficulties in obtaining a precise radiometric age for these and equivalent rocks.

(3) Either the base, or top, of the global glaciogenic deposits that mark this transitional period.

Other eras in the Archean and Proterozoic eons

Potential boundary type sections and GSSPs that provide more meaningful rock-based stratigraphic boundaries are being considered for other erathems. Potential *Paleoarchean–Mesoarchean* boundary sections occur in South Africa and in the Pilbara of Western Australia. Stratigraphic sections that record the onset of the late Archean super-event at about 2.78 Ga may be used to mark the *Mesoarchean–Neoarchean* boundary.

Despite the conceptual elegance of the Proterozoic periods, there has been little usage of this terminology by Precambrian workers because the periods did not practically reflect significant intervals in Earth history. The current *Paleoproterozoic–Mesoproterozoic* boundary at ~1.6 Ga appears to be ~100 myr too young to reflect accurately the peak of aggregation of the supercontinent Nuna, and should therefore be modified if a geodynamical concept is retained for that Proterozoic division.

The Geon concept

Alongside a chronometric and/or chronostratigraphic subdivision of the Precambrian, many researchers are using the Geon concept, which divides the Precambrian into chronometric divisions based on 100-myr intervals (e.g., Geon 34 = the period of time from 3.4 to 3.5 Ga). Although practical in terms of

Figure 3.3. The base of the Ediacaran System (GSSP) is defined as the base of the Marinoan cap carbonate (Nuccaleena Formation) in the Enorama Creek Section of the central Flinders Ranges, Adelaide Rift Complex, South Australia. The principal observed correlation events are (1) the rapid decay of Marinoan ice sheets and onset of distinct cap carbonates throughout the world, and (2) the beginning of a distinctive pattern of secular changes in carbon isotopes. The GSSP is underlain by a varied assemblage of glacial, glacial-marine, and associated deposits of the Elatina Formation. The lower contact of the Nuccaleena Formation is a disconformity, tentatively attributed to post-glacial isostatic rebound.

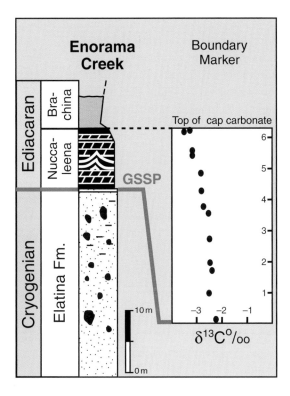

Figure 3.4. Stratigraphy of the base Ediacaran GSSP in the Enorama Creek Section of the central Flinders Ranges with the primary boundary markers.

global correlation of rock units, the Geon scheme is impractical for making geological maps and therefore cannot replace the eons and eras of the current stratigraphic chart.

Cryogenian and Ediacaran periods: the 250 myr before onset of the Cambrian

Widespread evidence of glacial deposits in the middle of the Neoproterozoic indicates at least two near-global glaciations. The first of the glacial episodes of this Cryogenian Period, called the "Sturtian" in Australia, occurred after 750 Ma and was over by 665 Ma. The later

Table 3.2 GSSP of Ediacaran stage, with location and primary correlation criteria

System/ Period	GSSP location	Latitude, longitude	Boundary level	Correlation events	Reference
Ediacaran	Enorama Creek, Flinders Ranges, South Australia	31° 19′ 53.2″ S 138° 38′ 0.2″ E	Base of the Marinoan cap carbonate	Beginning of a distinctive pattern of secular changes in carbon isotopes	*Lethaia* **39**, 2006

Source: Details are available at *www.stratigraphy.org* and in the *Episodes* publication.

"Marinoan" glaciation was of truly global extent and gave rise to the well-known epithet "Snowball Earth": it concludes the Cryogenian at ~635 Ma. The extent, cause, and effects of the Cryogenian glaciations are active topics of research and theoretical concepts, and we refer the reader to selected articles and authoritative websites listed at the end of this chapter for reviews.

While the Cryogenian Period will be defined to contain the main global ice ages of the Neoproterozoic, attempts to define a base for this period are concentrating on levels that might demonstrate a coincidence of biostratigraphic and geochemical events. Prior to the Sturtian ice age, there appears to have been an increase in complexity of microfossils, including the first suspected testate amoebae (vase-shaped microfossils), the first calci-microbes, a marked decrease in the diversity of stromatolitic-form genera, a significant negative carbon-isotope excursion, and the disappearance of "molar-tooth" microcrystalline calcite crack fill in carbonate rocks. The stratigraphic base of Neoproterozoic glaciogenic units is expected to vary in age from place to place, and thus is unsuitable for defining the base of the Cryogenian.

A texturally unusual carbonate deposit capping the Marinoan-glacial-derived deposits in many marine settings is associated with an exceptionally strong negative carbon-isotope excursion. These features were used to assign the GSSP for the base Ediacaran in the expanded Neoproterozoic–Cambrian succession within the Flinders Ranges of southern Australia in 2004 (Table 3.2).

The forthcoming global subdivision of the Ediacaran will be based on its stable isotope record, organic-walled microfossil biozones, Ediacaran metazoan fossils, and glacial and impact evidence. Magnetostratigraphy is being developed for intra- and interbasinal correlation.

Although the Ediacaran takes its name from the Ediacaran biota of dominantly soft-bodied benthic organisms, these earliest metazoans of uncertain affinity did not evolve until quite late in the Ediacaran Period: just after the short-lived Gaskiers glaciation at ~580 Ma and approximately coinciding with a major negative carbon-isotope anomaly (the "Wonoka" anomaly). The succession of the Ediacaran microfossils and metazoans enable a broad subdivision of the Ediacaran into four zones of the suggested Pertatataka, Avalon, Vendian, and Nama/Cloudina associations. The classic members of the Ediacaran biota disappear at the base of the Cambrian.

Acknowledgements

For further details/information, we recommend "The Precambrian: the Archean and Proterozoic Eons" by L. J. Robb, A. H. Knoll, K. A. Plumb, G. A. Shields, H. Strauss and J. Veizer

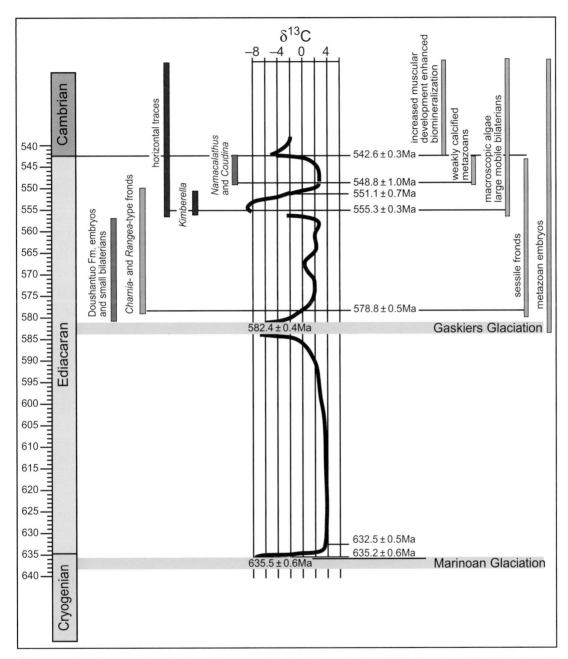

Figure 3.5. Carbon-isotope trends and major biotic events within the Ediacaran Period. [Modified from Fig. 2 of Daniel Condon *et al.*, *Science* 308: 95-98 (1 April 2005), which included data from Myrow and Kaufman, 1999. Used with permission from AAAS.]

and "Towards a 'natural' Precambrian time scale" by W. Bleeker (both in *A Geologic Time Scale 2004*). Portions of the background material are from documents of the Precambrian and Neoproterozoic subcommissions.

Further reading

Brocks, J. J., Buick, R., Summons, R. E., and Logan, G. A., 2003. A reconstruction of Archean biological diversity based on molecular fossils from the 2.78 to 2.45 billion-year-old Mount Bruce Supergroup, Hamersley Basin, Western Australia. *Geochimica et Cosmochimica Acta*, 22: 4321–4335.

Cavosie, A. J., Wilde, S. A., and Valley, J. W., 2005. A lower age limit for the Archean based on $\delta^{18}O$ of detrital zircons. *Geochimica et Cosmochimica Acta*, 69: A391.

Cloud, P., 1972. A working model of the primitive earth. *American Journal of Science*, 272: 537–548.

Condon, D., Zhu, M., Bowring, S., Wang, W., Yang, A., and Jin, Y., 2005. U–Pb ages from the Neoproterozoic Doushantuo Formation, China. *Science*, 308: 95–98.

Farquhar, J., Bao, H. M., and Thiemens, M., 2000. Atmospheric influence of Earth's earliest sulfur cycle. *Science*, 289: 756–758.

Fedo, C. M., Young, G. M., and Nesbitt, H. W., 1997. Paleoclimatic control on the composition of the Paleoproterozoic Serpent Formation, Huronian Supergroup, Canada: a greenhouse to icehouse transition. *Precambrian Research*, 86: 201–223.

Halverson, G. P., Hoffman, P. F., Schrag, D. P., Maloof, A. C., and Rice, A. H. N., 2005. Towards a Neoproterozoic composite carbon isotope record. *Geological Society of America Bulletin*, 117, 1181–1207.

Hoffmann, P. F., and Schrag, D. P., 2002. The snowball Earth hypothesis: testing the limits of global change. *Terra Nova*, 14: 129–155.

Kamber, S. B., 2007. The enigma of the terrestrial protocrust: evidence for its former existence and the importance of its complete disappearance. In: *Earth's Oldest Rocks*, eds. M. J. Van Kranendonk, R. H. Smithies, and V. Bennet. *Developments in Precambrian Geology*, 15: 75–90.

Knoll, A. H., 2003. *Life on a Young Planet: The First Three Billion Years of Evolution on Earth*. Princeton: Princeton University Press.

Knoll, A. H., Walter, M. R., Narbonne, G. M., and Christie-Blick, N., 2006. The Ediacaran Period: a new addition to the geologic time scale. *Lethaia*, 39: 13–30.

McCall, G. J. H., 2006. The Vendian (Ediacaran) in the geological record: enigmas in geology's prelude to the Cambrian explosion. *Earth Science Reviews*, 77: 1–229.

Melezhik, V. A., Fallick, A. E., Hanski, E. J., Kump, L. R., Lepland, A., Prave, A. R., and

Strauss, H., 2005. Emergence of the aerobic biosphere during the Archean–Proterozoic transition: challenges of future research. *GSA Today*, **15**: 4–11.

Papineau, D., Mojzsis, S. J., and Schmitt, A. K., 2007. Multiple sulphur isotopes from Paleoproterozoic Huronian interglacial sediments and the rise of atmospheric oxygen. *Earth and Planetary Science Letters*, **255**: 188–212.

Plumb, K. A., 1991. New Precambrian time scale. *Episodes*, **14**: 139–140.

Ryder, G., 2000. Planetary science: glass beads tell a tale of lunar bombardment. *Science*, **287**: 1768–1769.

Siebert, C., Kramers, J. D., Meisel, Th., Morel, Ph., and Nägler, Th. F., 2005. PGE, Re–Os, and Mo isotope systematics in Archean and early Proterozoic sedimentary systems as proxies for redox conditions of the early Earth. *Geochimica et Cosmochimica Acta*, **69**: 1787–1801.

Stern, R. A., and Bleeker, W., 1998. Age of the world's oldest rocks refined using Canada's SHRIMP: the Acasta Gneiss Complex, Northwest Territories. *Geoscience Canada*, **25**: 27–31.

Wilde, S. A., Valley, J. W., Peck, W. H., and Graham, C. M., 2001. Evidence from detrital zircons for the existence of continental crust and oceans on the Earth 4.4 Gyr ago. *Nature*, **409**: 175–178.

Zahnle, K., Arndt, N., Cockell, C., Halliday, A., Nisbet, E., Selsis, F., and Sleep, N. H., 2007. Emergence of a habitable planet. *Space Science Reviews*, **129**: 35–78. [A fascinating examination of possible events during the Hadean.]

Selected on-line references

Precambrian and Neoproterozoic subcommissions – *www.stratigraphy.org/ precambrian/* and *www.stratigraphy.org/ ediacaran/* (includes links to other sites)

Palaeos pages on Precambrian (Hadean, Archean, Proterozoic), especially the extensive summaries – *www.palaeos.com/ Proterozoic/Neoproterozoic/Cryogenian/ Cryogenian.html* and *Neoproterozoic/ Ediacaran/Ediacaran.html*

Planetary Habitability "web book" by Norman H. Sleep (Stanford University) – *pangea.stanford.edu/courses/gp025/webbook. html* [*a summary and speculations of Earth during the Hadean; prepared in a delightful style*]

Snowball Earth (Paul F. Hoffmann's site, funded by NSF) – *www.snowballearth.org*

We recommend the extensive Precambrian webpages and links at Smithsonian Institution, University of California Museum of Paleontology, and *Wikipedia*:

Wikipedia on-line encyclopedia (a public effort) – *en.wikipedia.org/wiki/Precambrian*, and links from that main page.

Smithsonian Institution paleobiology site – *paleobiology.si.edu/geotime/main/htmlVersion/*

archean1.html and *paleobiology.si.edu/geotime/ main/htmlVersion/proterozoic1.html.*

Authors

Martin van Kranendonk, Geological Survey of Western Australia, Mineral House, 100 Plain Street, East Perth, Western Australia 6004, Australia

James Gehling, South Australian Museum, North Terrace, Adelaide, South Australia 5000, Australia

Graham Shields, Geologisch-Paläontologisches Institut, Westfälische Wilhelms-Universität, 48149 Münster, Germany

4
Cambrian Period

Shanchi Peng and Loren Babcock

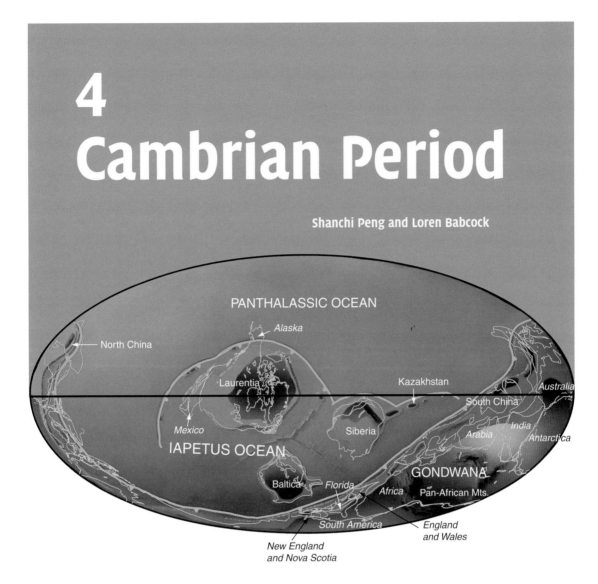

PANTHALASSIC OCEAN

Alaska

North China

Laurentia

Kazakhstan

Australia

South China

Mexico

Siberia

India

Arabia

Antarctica

IAPETUS OCEAN

GONDWANA

Baltica Florida

Africa Pan-African Mts.

South America

England and Wales

New England and Nova Scotia

Figure 4.1. Geographic distribution of the continents during the Cambrian Period (514 Ma). The paleogeographic map was provided by Christopher Scotese.

History and base of Cambrian (base of Phanerozoic)

The Cambrian is characterized by the appearance of mineralized skeletons of multicellular animals. The original Cambrian of Adam Sedgwick (Sedgwick and Murchison, 1835; Sedgwick, 1852) was based on lithostratigraphy of Wales without consideration of fossil content, and was named from *Cambria*, the Roman variant of the Celtic name *Cumbria* for Wales. The upper half of the current Cambrian system is essentially equivalent to Sedgwick's Lower Cambrian.

The Neoproterozoic–Cambrian boundary is part of one of the "greatest enigmas of the fossil record; i.e., the relatively abrupt appearance of skeletal fossils and complex, deep burrows in sedimentary successions around the world" (Brasier *et al.*,

Figure 4.2. The GSSP marking the base of the Cambrian System, and its lowermost Terreneuvian Series and Fortunian Stage, Fortune Head section, Burin Peninsula, Newfoundland, Canada.

1994). Until the late 1940s, the base of the Cambrian was generally placed at the lowest occurrence of trilobites. The discovery within underlying deposits of small shelly fossils (SSFs) in shallow carbonate facies and of distinctive burrows and other trace fossils in widespread siliciclastic facies led to the decision to locate the Neoproterozoic–Cambrian boundary near the onset of the *Trichophycus* (formerly called *Phycodes*) *pedum* trace fossil assemblage that reflects the appearance of complex sediment-disturbing behavior by multiple metazoans. This level is within the onset of a large negative basal-Cambrian carbon-isotope excursion (BACE) and the extinction of most Ediacaran-type organisms.

The low cliffs of Fortune Head in Newfoundland were chosen in 1992 as the GSSP site for the base of the Cambrian System. The global *Fortunian* Stage is named for this location, and the *Terreneuvian* Series is derived from Terre Neuve, the modern French name for Newfoundland.

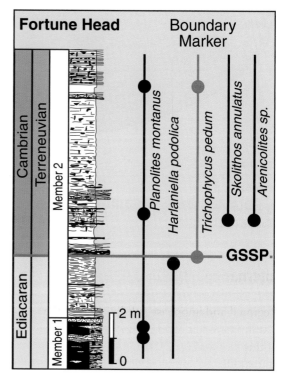

Figure 4.3. Stratigraphy of the base Cambrian GSSP in the Fortune Head section, Newfoundland, Canada with the primary boundary markers.

Table 4.1 GSSPs of Cambrian stages, with location and primary correlation criteria (status in 2008)

Stage	GSSP location	Latitude, Longitude	Boundary level	Correlation events	Reference
Stage 10	Candidate section is Duibian (Zhejiang Province, China)			Trilobite FAD of *Lotagnostus americanus*. An internal substage division might be FAD of *Codylodus adesei* conodont	
Stage 9	Candidate sections at Duibian (Zhejiang Province, China) and Gonggiri (Korea)			Trilobite FAD of *Agnostotes orientalis*	
Paibian	Wuling Mountains, Huayuan County, NW Hunan Province, China	28° 23.37′ N 109° 31.54′ E	At 396 m in the Huaqiao Formation	Trilobite FAD *Glyptagnostus reticulatus*	*Lethaia* **37**, 2004
Guzhangian	Louyixi, Guzhang County, NW Hunan Province, S China	28° 43.20′ N 109° 57.88′ E	121.3 m above the base of the Huaqiao Formation	Trilobite FAD *Lejopyge laevigata*	
Drumian	Drum Mountains, Millard County, Utah, USA	39° 30.705′ N 112° 59.489′ W	At the base of a dark-gray thinly laminated calcisiltite layer, 62 m above the base of the Wheeler Formation	Trilobite FAD *Ptychagnostus atavus*	*Episodes* **30**(2), 2007
Stage 5	Candidate sections are Wuliu-Zengjiayan (east Guizhou, China) and Split Mountain (Nevada, USA)			Trilobite, potentially FAD of *Oryctocephalus indicus*	
Stage 4				Trilobite FAD of *Olenellus* or *Redlichia*	
Stage 3				Trilobites – their FAD	
Stage 2				Small Shelly Fossils, or Archaeocyathid species	
Fortunian (base Cambrian)	Fortune Head, SE Newfoundland, Canada	47° 4′ 34.47″ N 55° 49′ 51.71″ W[a]	2.4 m above the base of Member 2 in the Chapel Island Formation	*Trace fossil* FAD *Trichophycus pedum*	*Episodes* **17**(1/2), 1994

a. According to Google Earth.
Source: Details on each GSSP are available at *www.stratigraphy.org* and in the *Episodes* publications.

International subdivisions of Cambrian

Regional and imprecise subdivisions of the Cambrian System had inhibited interpretation of this fascinating chapter in the origin of animal life. The limits and divisions of the Cambrian had evolved separately among regions, but none of these regional schemes had global applicability.

The International Subcommission on Cambrian Stratigraphy has worked hard during the past two decades to achieve an intercontinentally consistent time-stratigraphic language by first identifying a suite of biologic and chemostratigraphic horizons that could serve as stratigraphic tie points (Geyer and Shergold, 2000). These correlations led to a suite of international stages and their grouping into series. In 2005, a framework of four global series and ten stages was adopted (Babcock *et al.*, 2005), and GSSPs have been ratified for approximately half of these international stages (as of early 2008).

Traditional regional versions of a "Lower" Cambrian series contained a lower pre-trilobite interval that spanned nearly half the duration of the total Cambrian. Therefore, Cambrian stratigraphers have decided to divide this traditional "Lower" series of Cambrian into a lowermost pre-trilobite Terreneuvian Series followed by a "Series 2" that begins near the lowest occurrence of trilobites (Peng *et al.*, 2006). "Series 3" is an expanded version of typical regional variants of "Middle" Cambrian, with its base close to the traditional "Lower–Middle" Cambrian boundary. A reduced version of the "Upper" Cambrian is the *Furongian* Series (derived from the location of its basal GSSP in Hunan Province, the Lotus or *Furong* State of China).

These four global series are being divided into ten international stages, of which four have been defined by basal GSSPs. The base of the lowermost Fortunian Stage was placed near the lowest occurrence of a distinctive trace fossil assemblage. A primary marker for the base of the overlying "Stage 2" is potentially the lowest SSFs, and the base of "Stage 3" will be near the lowest trilobite (superfamily Fallotaspidoidea). The base of Stage 4 will be near the lowest occurrence of a trilobite of either the Olenellinae or the Redlichiina subfamilies, and the base of Stage 5 will be a horizon near the lowest occurrence of the polymerid trilobite *Oryctocephalus indicus*.

The GSSPs of the stages in middle Series 3 through the Furongian Series will be assigned at levels corresponding to the lowest occurrences of distinctive agnostoid trilobites, whose swimming pelagic niche led to widespread intercontinental distributions. So far, all GSSPs for the upper two series of the Cambrian system have been defined in outer shelf to slope environments. The names of the corresponding international stages (*Drumian*, *Guzhangian* and *Paibian* as of early 2008) are derived from the geographical locations of these GSSPs.

Selected aspects of Cambrian stratigraphy

Biostratigraphy

The pre-trilobite Terreneuvian Series has less-precise regional biostratigraphies based on SSFs of uncertain affinity and on archaeocyaths, an extinct relative of the sponges. The *Cambrian evolutionary explosion*, characterized by an apparent experimentation with a bizarre variety of body plans, was most significant during Series 2 (e.g., the Chengjiang deposits of China: Hou *et al.*, 2004) and lower Series 3 (e.g., the Burgess Shale of Canada: Conway Morris, 1998). All modern invertebrate phyla, except one, had their origin during this interval.

Trilobites, the best-known group of extinct arthropods, enable a fine biostratigraphic division of continental shelf and platform deposits. They attained their highest peak in diversity during Series 3 and the early Furongian of the Cambrian, but then suffered during the latest Cambrian. Conodonts, an important biostratigraphic tool for Ordovician through Triassic strata, appear within the lower Cambrian but only become useful for biostratigraphy beginning in the Furongian Series.

Chemostratigraphy

The Cambrian experienced some of the largest excursions in carbon isotopes of the entire Phanerozoic, and these have become powerful means for global correlation. A set of acronyms has been assigned based on their coincidence with other biological events or regional stratigraphy (e.g., *SPICE* is an acronym from "Steptoean Positive Isotope-Carbon Excursion" after its first identification within that North American stage; and *CARE* is abbreviated from "Cambrian Arthropod Radiation isotope Excursion" which is associated with the appearance of a wide variety of arthropods). Even though many of these major carbon isotope excursions seem to coincide with important biotic events or sea-level changes, the cause-and-effect relationships remain speculative.

Numerical time scale (current status and future developments)

An age of 542 ± 1 Ma for the Neoproterozoic–Cambrian boundary is from ash beds bracketing the boundary interval in Oman. The Cambrian–Ordovician boundary is well constrained at about 488 Ma. Between these levels, the ages for Cambrian series and stage boundaries are not as well constrained by the few U–Pb ages, and these boundaries have been largely redefined relative to the preliminary versions used in GTS04. The base of Series 2 is slightly older than the 519 Ma age on earliest trilobite-bearing strata in Wales. The base of Series 3 is slightly younger than the 511 Ma age on the *Protolenus*- and *Ellipsocephalus*-bearing strata in New Brunswick. The base of the Furongian Series is approximately 499 Ma.

In contrast to the time-scale calibration of other Paleozoic systems, there has not yet been a comprehensive effort to compile a standardized composite standard based on inter-regional biostratigraphy. With the establishment of cosmopolitan trilobite horizons for high-resolution ties of regional stratigraphies, it should now be possible to develop such a composite for the upper half of the Cambrian, which in turn will enable correlation and scaling to radiometric ages.

Acknowledgements

For further details/information, we recommend "The Cambrian Period" by J. H. Shergold and R. A. Cooper (in *A Geologic Time Scale 2004*). Portions of the background material are from documents of the Cambrian Subcommission.

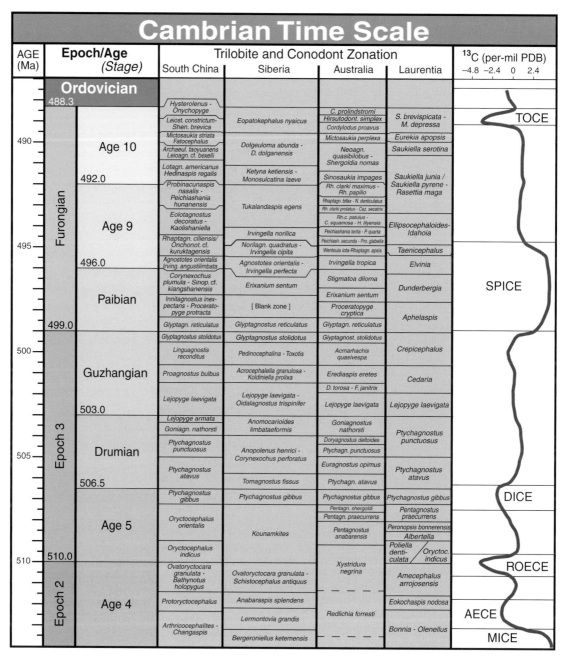

AGE (Ma)	Epoch/Age *(Stage)*		Trilobite and Conodont Zonation				^{13}C (per-mil PDB)
			South China	Siberia	Australia	Laurentia	−4.8 −2.4 0 2.4

Cambrian Time Scale

Ordovician

488.3

490	Furongian	Age 10	Hysterolenus - Onychopyge		C. prolindstromi	S. brevispicata - M. depressa	TOCE
			Leiost. constrictum- Shen. brevica	Eopatokephalus nyaicus	Hirsutodont. simplex		
					Cordylodus proavus		
			Mictosaukia striata Fatocephalus		Mictosaukia perplexa	Eurekia apopsis	
			Archaeul. taoyuanens Leioagn. cf. bexelli	Dolgeuloma abunda - D. dolganensis	Neoagn. quasibilobus - Shergoldia nomas	Saukiella serotina	
492.0			Lotagn. americanus Hedinaspis regalis	Ketyna ketiensis - Monosulcatina laeve	Sinosaukia impages	Saukiella junia / Saukiella pyrene - Rasettia maga	
		Age 9	Probinacunaspis nasalis - Peichiashania hunanensis		Rh. clarki maximus - Rh. papilio		
				Tukalandaspis egens	Rhaptagn. bifax - N. denticulatus		
					Rh. clarki prolatus - Caz. secatrix		
			Eolotagnostus decoratus - Kaolishaniella	Irvingella norilica	Rh.c. patulus - C. squamosa - H. lilyensis	Ellipsocephaloides- Idahoia	
495					Peichiashania tertia - P. quarta		
			Rhaptagn. ciliensis/ Onchonot. cf. kuruktagensis	Norilagn. quadratus - Irvingella cipita	Peichiash. secunda - Pro. glabella	Taenicephalus	
496.0					Wentsuia iota-Rhaptagn. apsis		
		Paibian	Agnostotes orientalis Irving. angustilimbata	Agnostotes orientalis - Irvingella perfecta	Irvingella tropica	Elvinia	SPICE
			Corynexochus plumula - Sinop. cf. kiangshanensis	Erixanium sentum	Stigmatoa diloma	Dunderbergia	
					Erixanium sentum		
			Innitagnostus inex- pectans - Procerato- pyge protracta	[Blank zone]	Proceratopyge cryptica	Aphelaspis	
499.0			Glyptagn. reticulatus	Glyptagnostus reticulatus	Glyptagnostus reticulatus		
500	Epoch 3	Guzhangian	Glyptagnostus stolidotus	Glyptagnostus stolidotus	Glyptagnost. stolidotus	Crepicephalus	
			Linguagnostis reconditus	Pedinocephalina - Toxotis	Acmarhachis quasivespa		
			Proagnostus bulbus	Acrocephalella granulosa - Koldiniella prolixa	Erediaspis eretes	Cedaria	
					D. torosa - F. janitrix		
			Lejopyge laevigata	Lejopyge laevigata - Oidalagnostus trispinifer	Lejopyge laevigata	Lejopyge laevigata	
503.0			Lejopyge armata	Anomocarioides limbataeformis	Goniagnostus nathorsti	Ptychagnostus punctuosus	
		Drumian	Goniagn. nathorsti		Doryagnostus deltoides		
505			Ptychagnostus punctuosus	Anopolenus henrici - Corynexochus perforatus	Ptychagn. punctuosus		
			Ptychagnostus atavus		Euragnostus opimus	Ptychagnostus atavus	
506.5				Tomagnostus fissus	Ptychagn. atavus		
		Age 5	Ptychagnostus gibbus	Ptychagnostus gibbus	Ptychagnostus gibbus	Ptychagnostus gibbus	DICE
	Epoch 2				Pentagn. shergoldi	Pentagnostus praecurrens	
					Pentagn. praecurrens		
			Oryctocephalus orientalis	Kounamkites	Pentagnostus anabarensis	Peronopsis bonnerensis	
						Albertella	
			Oryctocephalus indicus			Poliella denti- / Oryctoc. culata / indicus	
510.0			Ovatoryctocara granulata - Bathynotus holopygus	Ovatoryctocara granulata - Schistocephalus antiquus	Xystridura negrina	Amecephalus arrojosensis	ROECE
		Age 4	Protoryctocephalus	Anabaraspis splendens		Eokochaspis nodosa	
				Lermontovia grandis	Redlichia forresti		AECE
			Arthricocephalites - Changaspis	Bergeroniellus ketemensis		Bonnia - Olenellus	MICE

Figure 4.4. Numerical ages of epoch/series and age/stage boundaries of the Cambrian with major trilobite, archaeocyathid, and shelly fossil regional zonations and ^{13}C isotope fluctuations. ["Age" is the term for the time equivalent of the rock-record "stage".] Conodonts are used for some zones in the uppermost Cambrian. The ^{13}C isotope curve and events are from Zhu *et al.* (2006).

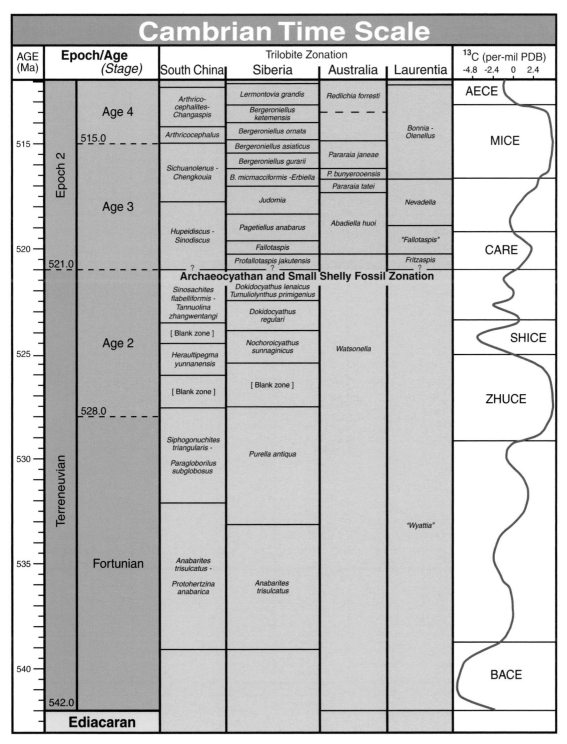

Figure 4.4 (cont.)

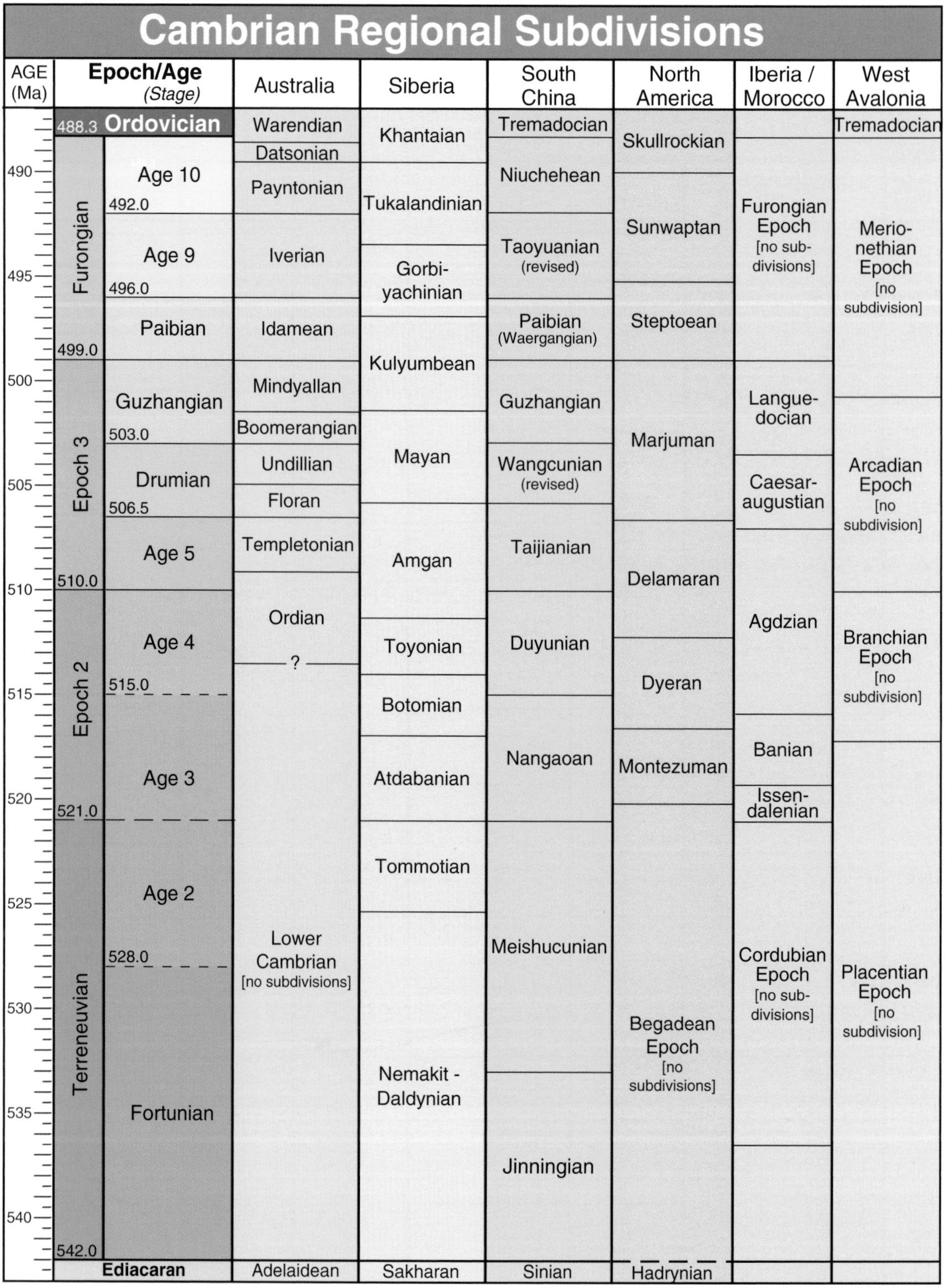

Cambrian Regional Subdivisions

AGE (Ma)	Epoch/Age (Stage)		Australia	Siberia	South China	North America	Iberia / Morocco	West Avalonia
488.3		Ordovician	Warendian	Khantaian	Tremadocian	Skullrockian		Tremadocian
490	Furongian	Age 10 / 492.0	Datsonian		Niuchehean		Furongian Epoch [no sub-divisions]	Merio-nethian Epoch [no subdivision]
			Payntonian	Tukalandinian		Sunwaptan		
495		Age 9 / 496.0	Iverian	Gorbi-yachinian	Taoyuanian (revised)			
		Paibian / 499.0	Idamean	Kulyumbean	Paibian (Waergangian)	Steptoean		
500	Epoch 3	Guzhangian / 503.0	Mindyallan		Guzhangian	Marjuman	Langue-docian	
			Boomerangian	Mayan				Arcadian Epoch [no subdivision]
505		Drumian / 506.5	Undillian		Wangcunian (revised)		Caesar-augustian	
			Floran					
		Age 5 / 510.0	Templetonian	Amgan	Taijianian	Delamaran		
510	Epoch 2	Age 4 / 515.0	Ordian — ? —	Toyonian	Duyunian		Agdzian	Branchian Epoch [no subdivision]
515				Botomian		Dyeran		
		Age 3 / 521.0		Atdabanian	Nangaoan	Montezuman	Banian	
520			Lower Cambrian [no subdivisions]				Issen-dalenian	
525	Terreneuvian	Age 2 / 528.0		Tommotian				
530				Meishucunian		Cordubian Epoch [no sub-divisions]	Placentian Epoch [no subdivision]
		Fortunian		Nemakit - Daldynian		Begadean Epoch [no subdivisions]		
535					Jinningian			
540								
542.0		Ediacaran	Adelaidean	Sakharan	Sinian	Hadrynian		

Figure 4.5. Correlation of the international subdivisions of the Cambrian System with selected regional stage nomenclatures. South China modified from Peng (2003) and Iberia/Morocco from Geyer and Landing (2004).

Further reading

Babcock, L. E., Peng, S. C., Geyer, G., and Shergold, J. H., 2005. Changing perspective on Cambrian chronostratigraphy and progress toward subdivision of the Cambrian System. *Geosciences Journal*, 9: 101–106.

Brasier, M. D., Cowie, J., and Taylor, M., 1994. Decision on the Precambrian–Cambrian boundary stratotype. *Episodes*, 17(1,2): 3–8.

Conway Morris, S., 1998. *The Crucible of Creation: The Burgess Shale and the Rise of Animals*. New York: Oxford University Press.

Geyer, G., and Landing, E., 2004. A unified Lower–Middle Cambrian chronostratigraphy for West Gondwana. *Acta Geologica Polonica* 54(2): 179–218.

Geyer, G., and Shergold, J., 2000. The quest for internationally recognized divisions of Cambrian time. *Episodes*, 23(3): 188–195.

Hou, X.-G., Aldridge, R. J., Bergstrom, J., Siveter, D. J., Siveter, D. J., and Feng, X.-H., 2004. *The Cambrian Fossils of Chengjiang, China: The Flowering of Early Animal Life*. London: Blackwell.

Landing, E., 1994. Precambrian–Cambrian boundary global stratotype ratified and a new perspective of Cambrian time. *Geology*, 22: 179–182.

Peng, S. C., 2003. Chronostratigraphic subdivision of the Cambrian of China. *Geologica Acta* 1: 135–144.

Peng, S. C., Babcock, L. E., Geyer, G., and Moczydłowska, M., 2006. Nomenclature of Cambrian epochs and series based on GSSPs: comments on an alternative proposal by Rowland and Hicks. *Episodes*, 29(2): 130–132.

Sedgwick, A., 1852. On the classification and nomenclature of the Lower Paleozoic rocks of England and Wales. *Quarterly Journal of the Geological Society of London* 8: 136–168.

Sedgwick, A., and Murchison, R. I., 1835. On the Cambrian and Silurian Systems exhibiting the order in which the older sedimentary strata succeeded each other in England and Wales. *London and Edinburgh Philosophical Magazine* 7: 483–485.

Zhu, M.-Y., Babcock, L. E., and Peng, S. C., 2006. Advances in Cambrian stratigraphy and paleontology: integrating correlation techniques, paleobiology, taphonomy and paleoenvironmental reconstruction. *Palaeoworld*, 15: 217–222.

Selected on-line references

Cambrian Subcommission – *www. palaeontologie.uni-wuerzburg.de/Stuff/casu6. htm* – Details on Cambrian stratigraphy, GSSPs and other aspects, including extensive bibliographies.

Virtual Cambrian (*wwwalt.uni-wuerzburg.de/ palaeontologie/Stuff/casu5.htm*) = under International Subcommission on Cambrian Stratigraphy.

A Guide to the Orders of Trilobites – *www. trilobites.info/* – An award-winning website devoted to understanding trilobites created and maintained by Sam Gon III.

Peripatus Paleontology "Cambrian Period" – *www.peripatus.gen.nz/Paleontology/ Ordovician.html* – amateur site, but quite extensive with additional Cambrian, Burgess Shale, and Vendian–Cambrian boundary links.

We recommend the extensive Cambrian webpages and links at *Palaeos*, Smithsonian Institution, University of California Museum of Paleontology, and *Wikipedia*. See URL details at end of Chapter 1.

Authors

Shanchi Peng, Nanjing Institute of Geology and Palaeontology, The Chinese Academy of Sciences, 39 East Beijing Street, Nanjing 210008, China

Loren E. Babcock, School of Earth Sciences, 125 South Oval Mall, The Ohio State University, Columbus, OH 43210, USA

5
Ordovician Period

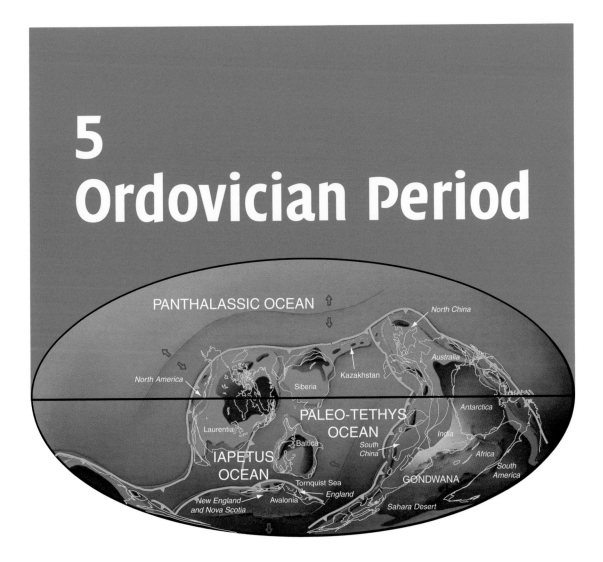

History and base of Ordovician

Figure 5.1. Geographic distribution of the continents during the Ordovician Period (485 Ma). The paleogeographic map was provided by Christopher Scotese.

The Ordovician System, named after the *Ordovices* tribe of Wales, was proposed by Charles Lapworth in 1879 to solve the controversy caused by the overlapping of the upper Cambrian of Adam Sedgwick and the lower Silurian of Roderick Murchison. The Ordovician was later extended downward to include the Tremadocian, but the current Ordovician was not officially accepted as an international unit until the International Geological Congress in 1960.

  Black graptolite-bearing shales are widely developed in Ordovician and Silurian basin facies throughout the world, and these graptolite successions have provided the primary method for global correlation and composite-scaling of Ordovician–Silurian events. The Ordovician essentially begins with the origin of this extinct group of floating colonial animals. The first occurrence of planktonic graptolites is slightly below the lowest occurrence of the conodont genus *Iapetognathus* of the cordylodid group. The Cambrian–Ordovician boundary

Figure 5.2. The GSSP marking the base of the Ordovician System and its lowermost Tremadocian Stage at Green Point, Newfoundland, Canada. The section is overturned to the west, thus the Ordovician beds are underneath the Cambrian. Photo provided by Stan Finney (former chair of Ordovician Subcommission).

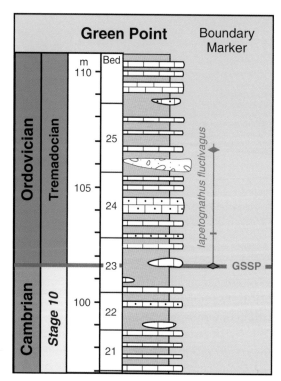

Figure 5.3. Stratigraphy of the base Ordovician GSSP in the Green Point section, Newfoundland, Canada, with the primary boundary markers.

GSSP on the Green Point coastal platform in western Newfoundland was placed to correspond to the lowest occurrence of the earliest *Iapetognathus* conodont species, rather than the lowest graptolite, because conodonts provided a better means of correlating between shelf and deep-water sequences. The Tremadocian Stage of lowest Ordovician is almost exactly coincident with the historical British Tremadoc series, and encompasses the interval during which planktonic graptolites became established as a major component of the oceanic macroplankton.

International subdivisions of Ordovician

Above the basal Tremadocian Stage, the traditional British subdivisions of Ordovician proved to be difficult to utilize for precise global correlation. Indeed, no single set of regional Ordovician units could be identified as adequate to serve as a global standard.

Table 5.1 GSSPs of Ordovician stages, with location and primary correlation criteria

Stage	GSSP location	Latitude, longitude	Boundary level	Correlation events	Reference
Hirnantian	Wangjiawan North section, N of Yichang city, Western Hubei Province, China	30° 59′ 2.68″ N 111° 25′ 10.76″ E	0.39 m below the base of the Kuanyinchiao Bed	Graptolite FAD *Normalograptus extraordinarius*	*Episodes* **29** (3), 2006
Katian	Black Knob Ridge Section, Atoka, Oklahoma, USA	34° 25.829′ N 96° 4.473′ W	4.0 m above the base of the Bigfork Chert	Graptolite FAD *Diplacanthograptus caudatus*	*Episodes* **30** (4), 2007
Sandbian	Sularp Brook, Fågelsång, Sweden	55° 42′ 49.3″ N 13° 19′ 31.8″ E[a]	1.4 m below a phosphorite marker bed in the E14b outcrop	Graptolite FAD *Nemagraptus gracilis*	*Episodes* **23** (2), 2000
Darriwilian	Huangnitang section, Changshan, Zhejiang Province, SE China	28° 51′ 14″ N 118° 29′ 23″ E[a]	Base of Bed AEP 184	Graptolite FAD *Undulograptus austrodentatus*	*Episodes* **20** (3), 1997
Dapingian	Huanghuachang section, NE of Yichang city, Hubei Province, China	30° 51′ 37.8″ N 110° 22′ 26.5″ E	10.57 m above base of the Dawan Formation	Conodont FAD of *Baltoniodus triangularis*	Proposal in *Episodes* **28** (2), 2005
Floian	Diabasbrottet, Hunneberg, Sweden	58° 21′ 32.2″ N 12° 30′ 08.6″ E	In the lower Tøyen Shale, 2.1 m above the top of the Cambrian	Graptolite FAD *Tetragraptus approximatus*	*Episodes* **27** (4), 2004
Tremadocian (base Ordovician)	Green Point Section, western Newfoundland	49° 40′ 58.5″ N 57° 57′ 55.09″ W[a]	At the 101.8-m level, within Bed 23, in the measured section	Conodont FAD *Iapetognathus fluctivagus*	*Episodes* **24** (1), 2001

a. According to Google Earth.
Source: Details on each GSSP are available at *www.stratigraphy.org* and in the *Episodes* publications.

Therefore, the international teams in the Subcommission on Ordovician Stratigraphy resolved to identify the best horizons for inter-regional correlation and subdivision. This framework enabled the establishment of a suite of stages and series in which the boundaries are GSSPs with global correlation potential. The Darriwilian was named for the Australian regional stage that spanned the same biostratigraphic interval, and the uppermost Hirnantian Stage closely corresponds to the same British stage. The other international stages are new names derived from geographic features near the GSSPs. This set of global stages of the Ordovician was completed in 2007 with the naming of the Dapingian Stage.

Graptolites served as the primary correlation horizons that guided placements of these GSSPs. Their global distribution is reflected in the widespread distribution of the GSSPs – Canada, Sweden, China, and USA. The only GSSP that does not have a graptolite as its primary correlation criteria is the Dapingian Stage, where the lowest occurrence of the conodont marker corresponds to major stratigraphic boundaries in other regions. Several of these graptolite-associated GSSPs were chosen for their coincidence with other significant events in Earth's history or eustatic rises of sea level. For example, the base of the Hirnantian is near the onset of the major glacial episode and a mass extinction in latest Ordovician.

The extensive research that enabled establishment of these international Ordovician stages also enabled an improved inter-regional correlation of regional stages, thereby enabling a greatly improved understanding of Ordovician history.

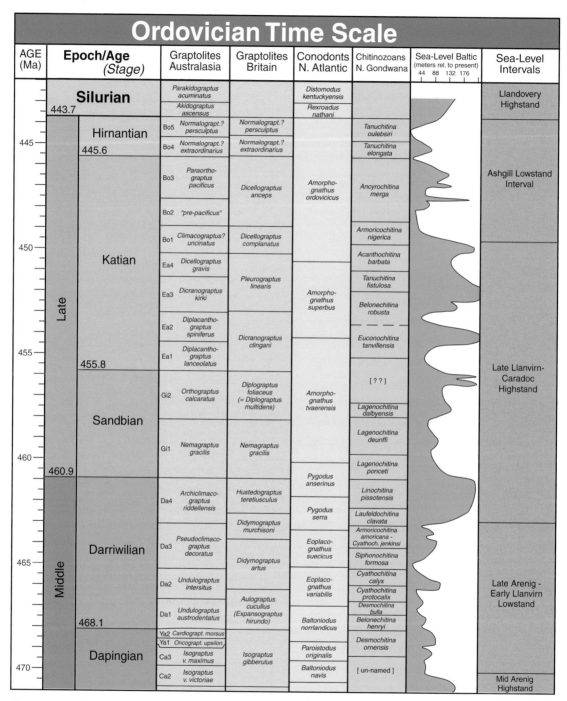

Ordovician Time Scale

AGE (Ma)	Epoch/Age (Stage)		Graptolites Australasia	Graptolites Britain	Conodonts N. Atlantic	Chitinozoans N. Gondwana	Sea-Level Baltic (meters rel. to present) 44 88 132 176	Sea-Level Intervals
443.7	**Silurian**		Parakidograptus acuminatus		Distomodus kentuckyensis			Llandovery Highstand
			Akidograptus ascensus		Rexroadus nathani			
445 445.6	Hirnantian	Bo5	Normalograpt.? persculptus	Normalograpt.? persculptus		Tanuchitina oulebsiri		
		Bo4	Normalograpt.? extraordinarius	Normalograpt.? extraordinarius		Tanuchitina elongata		Ashgill Lowstand Interval
	Katian	Bo3	Paraortho-graptus pacificus	Dicellograptus anceps	Amorpho-gnathus ordovicicus	Ancyrochitina merga		
		Bo2	"pre-pacificus"					
450		Bo1	Climacograptus? uncinatus	Dicellograptus complanatus		Armoricochitina nigerica		
		Ea4	Dicellograptus gravis			Acanthochitina barbata		
		Ea3	Dicranograptus kirki	Pleurograptus linearis	Amorpho-gnathus superbus	Tanuchitina fistulosa		
		Ea2	Diplacantho-graptus spiniferus	Dicranograptus clingani		Belonechitina robusta		
455 455.8		Ea1	Diplacantho-graptus lanceolatus			Euconochitina tanvillensis		Late Llanvirn-Caradoc Highstand
	Sandbian	Gi2	Orthograptus calcaratus	Diplograptus foliaceus (= Diplograptus multidens)	Amorpho-gnathus tvaerensis	[??]		
						Lagenochitina dalbyensis		
		Gi1	Nemagraptus gracilis	Nemagraptus gracilis		Lagenochitina deunffi		
460 460.9					Pygodus anserinus	Lagenochitina ponceti		
	Darriwilian	Da4	Archiclimaco-graptus riddellensis	Hustedograptus teretiusculus		Linochitina pissotensis		
				Didymograptus murchisoni	Pygodus serra	Laufeldochitina clavata		
		Da3	Pseudoclimaco-graptus decoratus	Didymograptus artus	Eoplaco-gnathus suecicus	Armoricochitina amoricana - Cyathoch. jenkinsi		
465						Siphonochitina formosa		Late Arenig - Early Llanvirn Lowstand
		Da2	Undulograptus intersitus		Eoplaco-gnathus variabilis	Cyathochitina calyx		
				Aulograptus cucullus (Expansograptus hirundo)		Cyathochitina protocalix		
468.1		Da1	Undulograptus austrodentatus		Baltoniodus norrlandicus	Desmochitina bulla		
						Belonechitina henryi		
	Dapingian	Ya2	Cardiograpt. morsus		Paroistodus originalis	Desmochitina ornensis		
		Ya1	Oncograpt. upsilon	Isograptus gibberulus				
		Ca3	Isograptus v. maximus		Baltoniodus navis	[un-named]		
470		Ca2	Isograptus v. victoriae					Mid Arenig Highstand

Figure 5.4. Numerical ages of epoch/series and age/stage boundaries of the Ordovician with major marine biostratigraphic zonations and sea-level changes. ["Age" is the term for the time equivalent of the rock-record "stage".] Biostratigraphic scales include graptolite, conodont, and chitinozoan zonations. The Australasian graptolite scale is from VandenBerg and Cooper (1992), Sadler and Cooper (2004), and John Laurie (pers. commun., May 2007). The British graptolite, conodont, and chitinozoan scales are from Webby *et al.* (2004). The sea-level curve and the sea-level intervals are from Nielsen (2004).

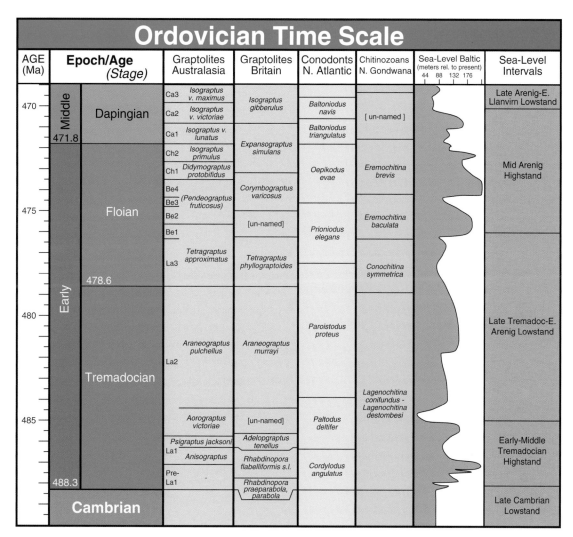

Ordovician Time Scale

AGE (Ma)	Epoch/Age *(Stage)*		Graptolites Australasia		Graptolites Britain	Conodonts N. Atlantic	Chitinozoans N. Gondwana	Sea-Level Baltic (meters rel. to present) 44 88 132 176	Sea-Level Intervals
470	Middle	Dapingian	Ca3	*Isograptus v. maximus*	*Isograptus gibberulus*	*Baltoniodus navis*	[un-named]		Late Arenig-E. Llanvirn Lowstand
			Ca2	*Isograptus v. victoriae*					
471.8			Ca1	*Isograptus v. lunatus*	*Expansograptus simulans*	*Baltoniodus triangulatus*			Mid Arenig Highstand
	Early	Floian	Ch2	*Isograptus primulus*		*Oepikodus evae*	*Eremochitina brevis*		
			Ch1	*Didymograptus protobifidus*					
475			Be4		*Corymbograptus varicosus*				
			Be3	*(Pendeograptus fruticosus)*		*Prioniodus elegans*	*Eremochitina baculata*		
			Be2		[un-named]				
			Be1						
478.6			La3	*Tetragraptus approximatus*	*Tetragraptus phyllograptoides*		*Conochitina symmetrica*		
480		Tremadocian	La2	*Araneograptus pulchellus*	*Araneograptus murrayi*	*Paroistodus proteus*			Late Tremadoc-E. Arenig Lowstand
485				*Aorograptus victoriae*	[un-named]	*Paltodus deltifer*	*Lagenochitina conifundus - Lagenochitina destombesi*		
			La1	*Psigraptus jacksoni*	*Adelopgraptus tenellus*				Early-Middle Tremadocian Highstand
				Anisograptus	*Rhabdinopora flabelliformis s.l.*	*Cordylodus angulatus*			
			Pre-La1	-	*Rhabdinopora praeparabola, parabola*				
488.3									
	Cambrian								Late Cambrian Lowstand

Figure 5.4. (cont.)

Selected aspects of Ordovician stratigraphy

Biostratigraphy

The *Great Ordovician Biodiversification Event* was a sustained radiation of diverse marine groups, and the level of biodiversity (about 1600 genera) was not significantly exceeded during the remainder of the Paleozoic.

Graptolites, conodonts, and chitinozoans are main inter-regional biostratigraphic groups. Trilobites, brachiopods, bryozoans, corals, ostracods, radiolarians, and nautiloid cephalopods provide regional or secondary zonations. Most groups suffered during the two phases of mass extinctions that accompanied the end-Ordovician glaciation and sea-level drops (earliest Hirnantian and latest Hirnantian events). This end-Ordovician climax is among the top five major extinction events in the geologic record.

Graptolites (extinct Phylum Hemicordata), which were particularly abundant in upwelling zones worldwide along continental margins, underwent major pulses of diversification during the Middle through early Late Ordovician. Their species span relatively brief durations, thereby enabling a high-resolution zonation. The most detailed zonal scheme is from Australasia, and a statistical compilation of these graptolite-rich sedimentary successions and correlation to other biogeographic regions was the basis for relative scaling of the stages within the Ordovician (and Silurian) Period for GTS04.

Conodonts, small eel-like animals resembling modern lampreys, only left their calcium phosphatic teeth as evidence of their rapidly evolving species. During the Ordovician, conodonts were in at least two major biogeographic provinces, and inter-regional correlations are uncertain in some intervals. Chitinozoans, organic-walled pelagic microfossils of uncertain affinity that first appeared in the Early Ordovician and began to diversify in the Middle Ordovician, are an increasingly useful addition to the well-developed graptolite and conodont tools. Trilobites, which had reached a moderate peak in diversity during Late Ordovician, suffered a near-extinction during the Hirnantian, after which they never regained their former prominence in Paleozoic marine fauna.

Sea level and chemostratigraphy

The Ordovician is considered to have the highest average sea level of the Phanerozoic, and marine deposits were emplaced far into the interiors of most continents. Major excursions, perhaps up to 250 m in magnitude, are superimposed onto this main trend. Major floodings (e.g., base of Late Ordovician) generally coincide with radiation of graptolites; and the preservation of black shales during these transgressions was probably enhanced by the generally warm greenhouse climate coupled with a lower oxygen level within the Ordovician atmosphere. A pronounced lowstand that accompanied the Hirnantian glacial and mass extinction episodes concluded the Ordovician.

Strontium isotopes ($^{87}Sr/^{86}Sr$) plunge across the Middle/Upper Ordovician boundary interval in the steepest sustained shift known from the Phanerozoic record. This extraordinary shift has been ascribed to the coincidence of enhanced mafic volcanism with a shutdown of cratonic weathering during the highstand peak. Carbon isotopes have two prominent peaks – a pronounced Hirnantian excursion and an earlier and lesser basal-Katian event – both considered to be indications of increased organic carbon burial, possible carbon dioxide drawdown, glaciation and sea-level lowstands.

Numerical time scale (GTS04 and future developments)

The Ordovician–Silurian time scale used in GTS04 is based on merging radiometric dates

Ordovician Regional Subdivisions

AGE (Ma)	Epoch/Age (Stage)	Britain	Australia	Baltoscandia	North America	China	Time Slices
443.7	**Silurian**		Keiloran	Juuru	Medinan		
445	Hirnantian 445.6 (Late)	Hirnantian (Ashgill)	Bolindian (Harju)	Porkuni	Gamachian (Cincinnatian)	Hirnantian	Hi2 / Hi1
	Katian (Late)	Rawtheyan (Ashgill)	Bolindian (Harju)	Pirgu	Rich-mondian (Cincinnatian)	Chientang-kiangian	Ka4
450	Katian	Cautleyan (Ashgill)	Bolindian	Pirgu	Rich-mondian	Chientang-kiangian	Ka4
	Katian	Pusgillian	Eastonian	Vormsi	Maysvillian		Ka3
	Katian	Streffordian (Caradoc)	Eastonian	Nabala	Edenian	Neichiasha-nian	Ka2
455	Katian 455.8	Cheneyan (Caradoc)	Eastonian	Rakvere / Oandu / Keila	Chat-fieldian (Mohawkian)	Neichiasha-nian	Ka1
	Sandbian	Burrellian (Caradoc)	Gisbornian	Haljala	Turinian	Neichiasha-nian	Sa2
460	Sandbian 460.9	Aurelucian (Caradoc)	Gisbornian (Viru)	Kukruse	Turinian		Sa1
	Darriwilian (Middle)	Llandeilian (Llanvirn)	Darriwillian	Uhaku / Lasnamagi / Aseri	White Rockian (Mohawkian)	Darriwilian	Da3
465	Darriwilian	Abereiddian (Llanvirn)	Darriwillian	Kunda	White Rockian	Darriwilian	Da2
	Darriwilian 468.1	Fennian (Arenig)	Yapeenian	Kunda	White Rockian	Darriwilian	Da1 / Dp3
470	Dapingian 471.8	Fennian (Arenig)	Castle-mainian	Volkhov	Rangerian (Ibexian)	Dapingian / Dawanian	Dp2 / Dp1
	Floian	Whitlandian (Arenig)	Chewtonian	Billingen	Black-hillsian (Ibexian)	Floian / Yushanian	Fl3 / Fl2
475	Floian	Moridunian (Arenig)	Bendi-gonian	Billingen (Oeland)	Black-hillsian	Floian / Yushanian	Fl1
480	Floian 478.6	Migneintian (Tremadoc)	Lance-fieldian	Hunne-berg	Tulean (Ibexian)	Tremadocian / Ichangian	Tr3
485	Tremadocian	Tremadocian (Tremadoc)	Lance-fieldian	Varangu	Stairsian	Tremadocian / Ichangian	Tr2
	Tremadocian 488.3	Tremadocian	Warendan	Pakerort	Skull-rockian		Tr1
	Cambrian		Datsonian				

Figure 5.5. Correlation of the international subdivisions of the Ordovician System with selected regional stage nomenclatures. British and North American stages are from Webby et al. (2004), Australian stages are provided by John Laurie (Geoscience Australia), and Chinese stages provided by Chen Xu. The British units are not the historical stages/series, but are a revised version (Webby et al., 2004). The time slices are from Bergström and Chen (2007).

with a global composite of graptolite events/ zones produced from constrained optimization (CONOP) correlation technique. The robust relative scaling of graptolite zones yielded the numerical ages for the graptolite-calibrated stage boundaries. Ages of all other biostratigraphic and chemostratigraphic events are derived from estimated calibrations to the graptolite composite.

Ordovician researchers have continued to improve the inter-calibration of fossil groups, geochemical proxies, and regional stratigraphies. The summary figures in this chapter include the final suite of international Ordovician stages and some of these post-GTS04 revisions in biostratigraphic, sea-level, and inter-regional correlations.

Acknowledgements

Chen Xu (Nanjing Institute of Geology and Palaeontology; chair of Ordovician Subcommission) contributed to this overview. For further details/information, we recommend "The Ordovician Period" by R. A. Cooper and P. M. Sadler (in *A Geologic Time Scale 2004*). Portions of the background material are from unpublished documents of the Ordovician Subcommission.

Further reading

Bergström, S. M., and Chen, X., 2007. Ordovician correlation chart of regional stages. [JPG posted in mid-2007 onto website of the Ordovician Subcommision (*www.ordovician.cn*), downloaded 3 July 2007.]

Finney, S., 2005. Global series and stages for the Ordovician System: a progress report. *Geologica Acta*, **3**: 309–316 (free access at *www.geologica-acta.com*).

Nielsen, A. T., 2004. Ordovician sea level changes: a Baltoscandian perspective. In: *The Great Ordovician Biodiversity Event*, eds. B. D. Webby, F. Paris, M. L. Droser, and I. G. Percival. New York: Columbia University Press, pp. 84–93.

Sadler, P. M., and Cooper, R. A., 2004. Calibration of the Ordovician time scale. In: *The Great Ordovician Biodiversity Event*, eds. B. D. Webby, F. Paris, M. L. Droser, and I. G. Percival. New York: Columbia University Press, pp. 48–51.

Saltzman, M. R., and Young, S. A., 2005. Long-lived glaciation in the Late Ordovician? Isotopic and sequence-stratigraphic evidence from western Laurentia. *Geology*, **33**: 109–112.

VandenBerg, A. H. M., and Cooper, R. A., 1992. The Ordovician graptolite sequence of Australasia. *Alcheringa*, **16**: 33–65.

Webby, B. D., Paris, F., Droser, M. L., and Percival, I. G. (eds.), 2004. *The Great Ordovician Biodiversity Event*. New York: Columbia University Press. [In addition to syntheses for most fossil groups, the book has excellent summaries of Ordovician climate, geochemistry and sea-level changes.]

Selected on-line references

Ordovician Subcommission – *www.ordovician. cn* – Details on Ordovician stratigraphy,

GSSPs, regional correlation charts, and other aspects, including extensive bibliographies.

Peripatus Paleontology "Ordovician Period" – *www.peripatus.gen.nz/Paleontology/ Ordovician.html* – an amateur site, but quite extensive with additional Ordovician links.

We recommend the extensive Ordovician webpages and links at *Palaeos*, Smithsonian Institution, University of California Museum of Paleontology, and *Wikipedia*. See URL details at end of Chapter 1.

6 Silurian Period

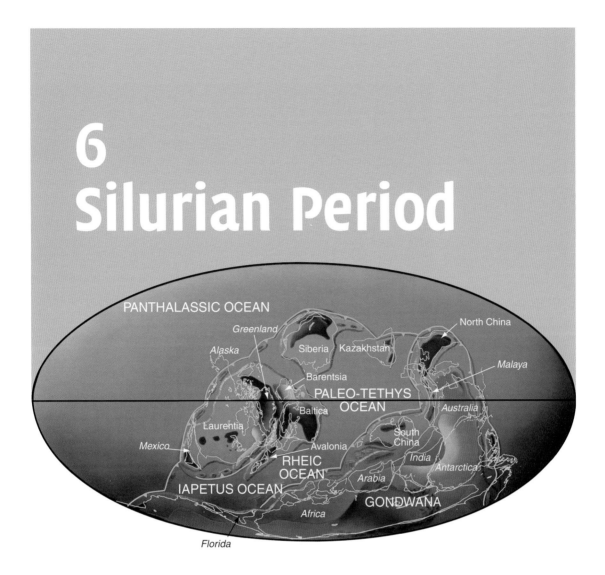

PANTHALASSIC OCEAN

Greenland
Alaska
Siberia Kazakhstan
North China
Barentsia
PALEO-TETHYS OCEAN
Malaya
Baltica
Laurentia
Australia
Mexico
Avalonia
South China
India
RHEIC OCEAN
Antarctica
Arabia
IAPETUS OCEAN
GONDWANA
Africa
Florida

History and base of Silurian

The Silurian System, named after the *Silures* tribe of Wales, was erected by Roderick Murchison in 1839. The current Silurian System corresponds to the upper portion of Murchison's version.

The Ordovician–Silurian boundary in the graptolite-bearing shales of Dob's Linn in Scotland marks the initial stages of recovery from the end-Ordovician mass extinctions. This base-Silurian GSSP coincides with the

Figure 6.1. Geographic distribution of the continents during the Silurian Period (425 Ma). The paleogeographic map was provided by Christopher Scotese.

first appearance of the graptolite *Akidograptus ascensus*, defining the base of the *A. ascensus* Biozone. [Note: The previously published assignment as coincident with the local base of the *Parakidograptus acuminatus* Zone is now known to be incorrect, and this had caused conflicting and confusing correlations.]

Figure 6.2. The GSSP marking the base of the Silurian System and its lowermost Rhuddanian Stage, at Dob's Linn, Scotland. The section is overturned. The level of the GSSP is marked in red, the base of the Birkhill Shale is marked by the yellow meter stick, and the top of the *Parakidograptus acuminatus* graptolite Zone is approximately at the position of the backpack on the lower left. Photo by Michael Melchin.

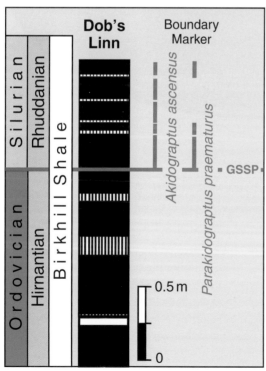

Figure 6.3. Stratigraphy of the base-Silurian GSSP in the Dob's Linn section, Scotland, with the primary boundary markers.

International subdivisions of Silurian

The Silurian underwent a complex history of subdivisions, definitions and nomenclature until the current suite of stages and series was formalized by the Silurian Subcommission in 1984. The classical Wales–England borderland region used by Murchison to establish the Silurian is the primary source for the nomenclature and boundary stratotypes for the three main series of the Silurian – Llandovery, Wenlock, and Ludlow – and their component stages. However, some of these stage GSSPs have been difficult to correlate outside of their local region. Even though this region has a diverse shallow-marine fossil content, there is often a lack of well-defined markers of the more cosmopolitan graptolite and conodont zonations. In addition, some of the GSSPs were placed at lithologic contacts that do not coincide with known widespread biostratigraphic horizons. For example, the

Table 6.1 GSSPs of Silurian stages, with location and primary correlation criteria

Stage	GSSP location	Latitude, longitude	Boundary level	Correlation events	Reference
Pridoli (Series)	Pozáry Section, Reporyje, Prague, Czech Republic	50° 01′ 39.82″ N 14° 19′ 29.56″ E[a]	Within Bed 96	Graptolite FAD Monograptus parultimus	Episodes **8**(2), 1985; Geological Series, National Museum of Wales **9**, 1989
Ludfordian	Near Ludlow, UK	52° 21′ 33″ N 2° 46′ 38″ W[a]	Coincident with the base of the Leintwardine Formation	Imprecise; may be near base of Saetograptus leintwardinensis graptolite zone	Lethaia **14**, 1981; Episodes **5**(3), 1982; Geological Series, National Museum of Wales **9**, 1989
Gorstian	Near Ludlow, UK	52° 21′ 33″ N 2° 46′ 38″ W[a]	Coincident with the base of the Lower Elton Formation	Graptolite FAD Saetograptus (Colonograptus) varians	Lethaia **14**, 1981; Episodes **5**(3), 1982; Geological Series, National Museum of Wales **9**, 1989
Homerian	Sheinton Brook, Homer, UK	52° 36′ 56″ N 2° 33′ 53″ W[a]	Within upper part of the Apedale Member of the Coalbrookdale Formation	Graptolite FAD Cyrtograptus lundgreni	Lethaia **14**, 1981; Episodes **5**(3), 1982; Geological Series, National Museum of Wales **9**, 1989
Sheinwoodian	Hughley Brook, UK	52° 34′ 52″ N 2° 38′ 20″ W[a]	Base of the Buildwas Formation	Imprecise. Between the base of acritarch biozone 5 and condont LAD of Pterospathodus amorphognathoides. The current GSSP does not coincide with the base of the Cyrtograptus centrifugus Biozone, as was supposed when the GSSP was defined. Restudy recommends a slightly higher and correlatable level on condonts – the Ireviken datum 2, which coincides approximately with the base of the murchisoni graptolite Biozone	Lethaia **14**, 1981; Episodes **5**(3), 1982; Geological Series, National Museum of Wales **9**, 1989
Telychian	Cefn-cerig Road Section, Wales, UK	51.97° N 3.79° W[b]	Within the Wormwood Formation	Just above Brachiopod LAD of Eocoelia intermedia and below Eocoelia FAD of curtisi	Episodes **8**(2), 1985; Geological Series, National Museum of Wales **9**, 1989
Aeronian	Trefawr Track Section, Wales, UK	52.03° N 3.70° W[b]	Within Trefawar Formation	Graptolite FAD Monograptus austerus sequens	Geological Series, National Museum of Wales **9**, 1989
Rhuddanian (base Silurian)	Dob's Linn, Scotland	55.44° N 3.27° W[b]	1.6 m above the base of the Birkhill Shale Formation	Graptolite FAD Akidograptus ascensus	Episodes **8**(2), 1985

a. According to Google Earth.
b. Derived from map.
Source: Details on each GSSP are available at *www.stratigraphy.org* and in Holland and Bassett (1989).

base-Wenlock GSSP (base of Sheinwoodian Stage) was placed at a prominent lithologic contact with no known correspondence to graptolite, conodont, or acritarch markers. As a result, nearly half of these GSSPs appear to be unsuitable for high-resolution global use. However, the current Silurian time scale assumes a correlation of stage boundaries with graptolite zones, and this proxy graptolite scheme will serve until the actual GSSPs are clarified or moved to more suitable locations.

In Britain, the "top of Silurian" had been traditionally been placed at a Ludlow Bone Bed below non-marine strata. However, the establishment of the Silurian–Devonian boundary near Prague in Czech Republic in 1972 (see next chapter) left an interval between the uppermost fossils in the British Silurian and the newly defined base of the international Devonian. This interval was given a separate series-rank status as the Prídolí, with its basal GSSP in graptolite-bearing platy limestone near the Silurian–Devonian boundary GSSP near Prague. However, the lack of stage-level subdivisions for the Prídolí makes this Series unique within the Phanerozoic chronostratigraphy.

Selected aspects of Silurian stratigraphy

Biostratigraphy

Graptolites, conodonts, and chitinozoans serve as the main biostratigraphic tools for inter-regional correlation. The Silurian Subcommission has compiled "generalized zonal sequences" for each of these groups, in which each zonal boundary can be identified in multiple paleogeographic realms. As in the Ordovician, the graptolites generally provide the primary standard and their high-resolution zones have an average duration of less than 1 million years. Major extinction events within the graptolites generally occur during Silurian sea-level regressions or lowstands.

Land plants began their evolution during the Silurian and the associated sporomorph zones provide a broad zonation. The evolution of shark-like fish left a distinctive succession of scales and other fossil parts that comprise a broad vertebrate biostratigraphic zonation.

Sea level and chemostratigraphy

Average Silurian sea level was lower than the Ordovician. In general, each Silurian stage contains one major oscillation of eustatic sea level. Based on onlap relations to paleo-valleys, the magnitude of the major oscillations exceeded 60 m. Positive excursions of oxygen-18 (lower Sheinwoodian, upper Homerian, and upper Ludfordian) occur during lowstand episodes, which have been interpreted as possible signatures of glaciation. However, there is not yet any unambiguous physical evidence to support these possible glaciation episodes. As in the Ordovician, positive excursions of carbon isotopes (increased carbon-13) seem to coincide with these same levels.

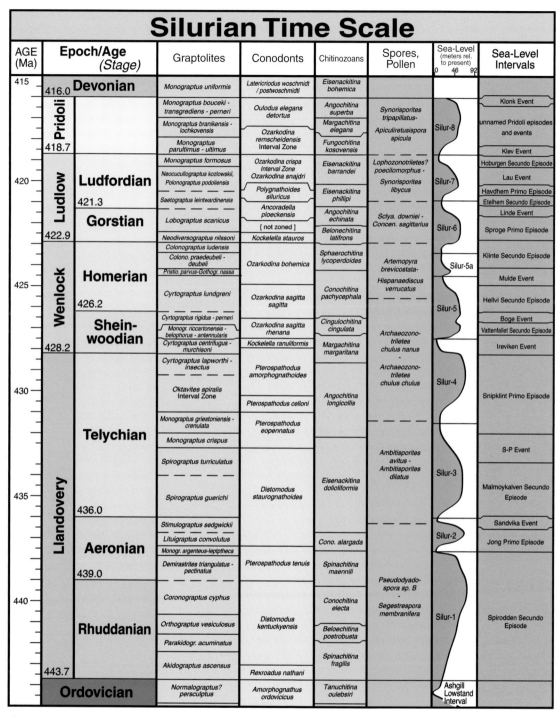

Silurian Time Scale

AGE (Ma)	Epoch/Age (Stage)		Graptolites	Conodonts	Chitinozoans	Spores, Pollen	Sea-Level (meters rel. to present) 0 46 92	Sea-Level Intervals
415 — 416.0		Devonian	Monograptus uniformis	Latericriodus woschmidt / postwoschmidti	Eisenackitina bohemica			
	Pridoli		Monograptus bouceki - transgrediens - perneri	Oulodus elegans detortus	Angochitina superba	Synorisporites tripapillatus-	Silur-8	Klonk Event
			Monograptus branikensis - lochkovensis	Ozarkodina remscheidensis Interval Zone	Margachitina elegans	Apiculiretusispora spicula		unnamed Pridoli episodes and events
418.7			Monograptus parultimus - ultimus		Fungochitina kosovensis			Klev Event
420 —	Ludlow	Ludfordian	Monograptus formosus	Ozarkodina crispa Interval Zone Ozarkodina snajdri	Eisenackitina barrandei	Lophozonotriletes? poecilomorphus -	Silur-7	Hoburgen Secundo Episode
			Neocucullograptus kozlowskii, Polonograptus podoliensis			Synorisporites libycus		Lau Event
421.3			Saetograptus leintwardinensis	Polygnathoides siluricus	Eisenackitina phillipi			Havdhem Primo Episode
				Ancoradella ploeckensis	Angochitina echinata	Sclya. downiei - Concen. sagittarius		Etelhem Secundo Episode
		Gorstian	Lobograptus scanicus	[not zoned]			Silur-6	Linde Event
422.9			Neodiversograptus nilssoni	Kockelella stauros	Belonechitina latifrons			Sproge Primo Episode
	Wenlock	Homerian	Colonograptus ludensis		Sphaerochitina lycoperdoides			Klinte Secundo Episode
			Colono. praedeubeli - deubeli	Ozarkodina bohemica		Artemopyra brevicostata-	Silur-5a	
			Pristio. parvus-Gothogr. nassa					Mulde Event
425 — 426.2			Cyrtograptus lundgreni	Ozarkodina sagitta sagitta	Conochitina pachycephala	Hispanaediscus verrucatus	Silur-5	Hellvi Secundo Episode
		Shein-woodian	Cyrtograptus rigidus - perneri	Ozarkodina sagitta rhenana	Cingulochitina cingulata			Boge Event
			Monogr. riccartonensis - belophorus - antennularis			Archaeozono-triletes chulus nanus		Vattenfallet Secundo Episode
428.2			Cyrtograptus centrifugus - murchisoni	Kockelella ranuliformis	Margachitina margaritana			Ireviken Event
	Llandovery	Telychian	Cyrtograptus lapworthi - insectus	Pterospathodus amorphognathoides	Angochitina longicollis	Archaeozono-triletes chulus chulus	Silur-4	Snipklint Primo Episode
430 —			Oktavites spiralis Interval Zone	Pterospathodus celloni				
			Monograptus griestoniensis - crenulata	Pterospathodus eopennatus				
			Monograptus crispus					S-P Event
			Spirograptus turriculatus		Eisenackitina dolioliformis	Ambitisporites avitus - Ambitisporites dilatus	Silur-3	
435 —			Spirograptus guerichi	Distomodus staurognathoides				Malmoykalven Secundo Episode
436.0			Stimulograptus sedgwickii					Sandvika Event
		Aeronian	Lituigraptus convolutus		Cono. alargada		Silur-2	Jong Primo Episode
			Monogr. argenteus-leptptheca					
439.0			Demirastrites triangulatus - pectinatus	Pterospathodus tenuis	Spinachitina maennili	Pseudodyado-spora sp. B - Segestrespora membranifera		
440 —			Coronograptus cyphus		Conochitina electa		Silur-1	Spirodden Secundo Episode
		Rhuddanian	Orthograptus vesiculosus	Distomodus kentuckyensis	Beloechitina postrobusta			
			Parakidogr. acuminatus		Spinachitina fragilis			
443.7			Akidograptus ascensus	Rexroadus nathani				
		Ordovician	Normalograptus? persculptus	Amorphognathus ordovicicus	Tanuchitina oulebsiri		Ashgill Lowstand Interval	

Figure 6.4. Numerical ages of epoch/series and age/stage boundaries of the Silurian with major marine biostratigraphic zonations and sea-level changes. ["Age" is the term for the time equivalent of the rock-record "stage".] The graptolite and chitinozoan scales are the standard zonations for the Silurian, the conodont scale is modified from Johnson (2006), and the spore-pollen scale is from Melchin *et al.* (GTS04). The sea-level curve is from Johnson (2006). The sea-level intervals are from Jeppsson (1998) as shown in Johnson (2006).

Strontium isotopes (^{87}Sr/^{86}Sr) display a slow rise through the Silurian that has been attributed to an increase in fluvial flux of radiogenic Sr due to climate warming and lowered sea level.

Numerical time scale (GTS04 and future developments)

As explained in the Ordovician chapter, the Ordovician–Silurian time scale used in GTS04 is based on a global composite of graptolite events/zones merged with radiometric dates.

The inter-calibration of Silurian biozones, geochemistry and eustatic events has attained a relative stability, thanks to the past efforts of Silurian workers and the Silurian Subcommission. The main aspects that remain relatively uncertain are correlation of terrestrial evolution and the history of the geomagnetic field.

Acknowledgements

Jiayu Rong (Nanjing Institute of Geology and Palaeontology; chair of Silurian Subcommission) and Michael Melchin (St. Francis Xavier University, Canada) contributed to this overview. For further details/information, we recommend "The Silurian Period" by M. J. Melchin, R. A. Cooper, and P. M. Sadler (in *A Geologic Time Scale 2004*). Portions of the background material are from unpublished documents of the Silurian Subcommission.

Further reading

Holland, C. H., and Bassett, M. G. (eds.), 1989. *A Global Standard for the Silurian System*, Geological Series No. 10. Cardiff: National Museum of Wales.

Jeppsson, L., 1998. Silurian oceanic events; summary of general characteristics. In: *Silurian Cycles: Linkages of Dynamic Stratigraphy with Atmosphere, Oceanic and Tectonic Changes*, eds. E. Landing and M. E. Johnson. *New York State Museum Bulletin*, **491**: 239–257.

Johnson, M. E., 2006. Relationship of Silurian sea-level fluctuations to oceanic episodes and events. *GFF* (journal formerly called *Geologiska Föreningens i Stockholm Förhandlingar*), **128**: 115–121.

Kaljo, D., and Martma, T., 2006. Application of carbon isotope stratigraphy to dating the Baltic Silurian rocks. *GFF* (journal formerly called *Geologiska Föreningens i Stockholm Förhandlingar*), **128**: 123–129

Landing, E., and Johnson, M. E. (eds.), 1998. *Silurian Cycles: Linkages of Dynamic Stratigraphy with Atmosphere, Oceanic and Tectonic Changes. New York State Museum Bulletin*, **491**: 1–327.

Selected on-line references

Silurian Subcommission – *www.silurian.cn/home.asp* – details on Silurian stratigraphy and GSSPs.

Virtual Silurian Reef site of Milwaukee Public Museum – *www.mpm.edu/collections/learn/reef/index.html*.

Peripatus Paleontology "Silurian Period" – *www.peripatus.gen.nz/Paleontology/Silurian.html* – amateur site, but quite extensive with additional Silurian links.

We recommend the extensive Silurian webpages and links at *Palaeos*, Smithsonian Institution, University of California Museum of Paleontology, and *Wikipedia*. See URL details at end of Chapter 1.

7
Devonian Period

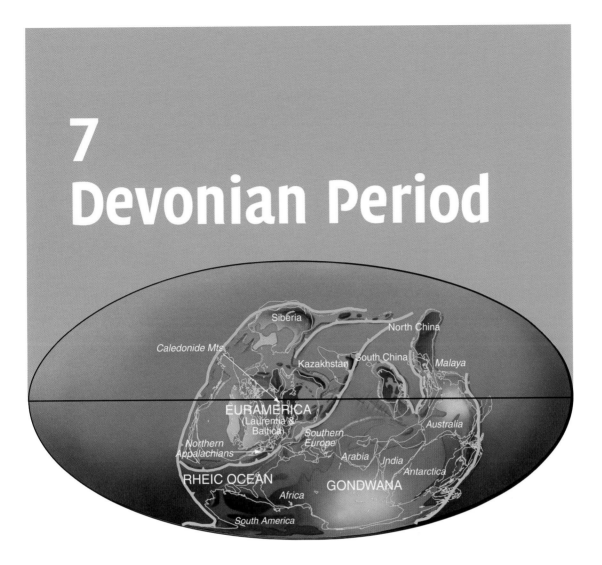

Figure 7.1. Geographic distribution of the continents during the Devonian Period (390 Ma). The paleogeographic map was provided by Christopher Scotese.

History and base of Devonian

The Devonian System was named after rock exposures in Devon county of England by Roderick Murchison and Adam Sedgwick in 1839. The collision of Baltica and Laurentia in the Caledonian orogeny resulted in the shedding of "Old Red Sandstone" across much of Laurasia. In Britain, the type Silurian is truncated by the non-marine Old Red Sandstone, therefore the "historical stratotype" region was unsuitable for defining the basal boundary and most Devonian stages.

Usage of the Silurian–Devonian boundary was inconsistent among continents, partly because it was assumed that the extinction of graptolites occurred at this level. The international agreement to place the GSSP for Silurian–Devonian boundary within a graptolite-bearing succession at Klonk, near Prague in the Czech Republic in 1972 resolved this problem. The base of the Devonian was assigned within the dark platy limestone

Figure 7.2. The GSSP marking the base of the Devonian System and its lowermost Lochkovian Stage, at Klonk, Czech Republic.

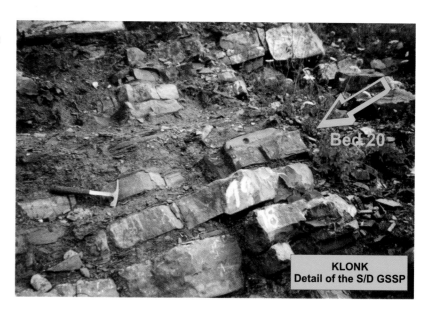

Figure 7.3. Stratigraphy of the base-Devonian GSSP in the section at Klonk, Czech Republic, with the primary boundary markers.

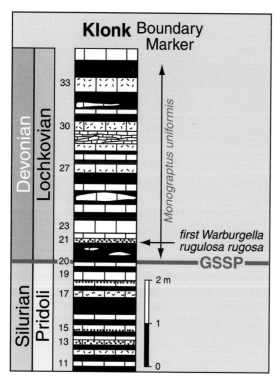

succession at a level just below the lowest occurrence of graptolite *Monograptus uniformis*. Recognition of the boundary in carbonate strata or in sections that lack graptolites is enabled by a major turnover in conodont species and a major positive carbon-isotope excursion at the anoxic Klonk Event that straddles the boundary interval. This boundary stratotype, ratified in 1972, has the distinction of being the first official GSSP. The published decision (1977) was accompanied by the establishment of principles for assigning GSSPs for global stages.

International subdivisions of Devonian

The suite of international stages was completed with the ratification of the Emsian in 1995. Except for the basal GSSP, which utilizes a graptolite as its primary correlation horizon, the GSSPs for all Devonian stages are associated with conodont events. The GSSP locations are in the Czech Republic, Germany, France and

Table 7.1 GSSPs of Devonian stages, with location and primary correlation criteria

Stage	GSSP location	Latitude, longitude	Boundary level	Correlation events	Reference
Famennian	Coumiac Quarry, near Cessenon, Montagne Noire, France	43° 27′ 40.6″ N 3° 02′ 25″ E[a]	Base of Bed 32a	Conodont FAD *Palmatolepis subperlobata* and flood occurrence of *Palmatolepis ultima*	*Episodes* **16**(4), 1993
Frasnian	Col du Puech de la Suque, Montagne Noire, France	43° 30′ 11.4″ N 3° 05′ 12.6″ E[a]	Base of Bed 42′ at Col du Puech de la Suque section E	Conodont FAD *Ancyrodella rotundiloba*	*Episodes* **10**(2), 1987
Givetian	Jebel Mech Irdane, Morocco	31° 14′ 14.7″ N 4° 21′ 14.8″ W[a]	Base of Bed 123	Conodont FAD *Polygnathus hemiansatus*	*Episodes* **18**(3), 1995
Eifelian	Wetteldorf, Eifel Hills, Germany	50° 08′ 58.6″ N 6° 28′ 17.6″ E[a]	21.25 m above the base of the exposed section, base of unit WP30	Conodont FAD *Polygnathus costatus partitus*	*Episodes* **8**(2), 1985
Emsian	Zinzil'ban Gorge, Uzbekistan	39° 12′ N 67° 18′ 20″ E	Base of Bed 9/5 in the Zinzil'ban Gorge in the Kitab State Geological Reserve	Conodont FAD *Polygnathus kitabicus*	*Episodes* **20**(4), 1997
Pragian	Velká Chuchle, Prague, Czech Republic	50° 00′ 53″ N 14° 22′ 21.5″ E[a]	Base of Bed 12 in Velká Chuchle Quarry	Conodont FAD *Eognathodus sulcatus sulcatus* or, better, of *Latericriodus steinachensis* Morph beta	*Episodes* **12**(2), 1989
Lochkovian (base Devonian)	Klonk, near Prague, Czech Republic	48.855° N 13.792° E[b]	Within Bed 20	Graptolite FAD *Monograptus uniformis*	IUGS, Series A, **5**, 1977

a. According to Google Earth.
b. Derived from map.
Source: Details on each GSSP are available at *www.stratigraphy.org* and in the *Episodes* publication.

Morocco. Even though difficulties in precise correlation from some GSSPs have led to requests for reconsideration, most of the GSSPs have secondary markers.

Selected aspects of Devonian stratigraphy

Biostratigraphy

Conodonts, the tooth and jaw elements of eel-like vertebrates, provide a standardized global biostratigraphic framework for the Devonian. Conodonts reached their greatest diversification during the Late Devonian. A standardized stratigraphy of conodonts is the primary scale for global correlation, placement of boundary GSSPs, and extrapolating numerical ages.

Graptolites, which provided the foundation for inter-regional correlation of Ordovician–Silurian marine strata, become extinct in the Early Devonian (earliest Emsian) simultaneously with an increase in the diversity and utility of ammonoids (goniatites and clymenids) for biostratigraphy. A standardized ammonite scale, largely based on characteristic genera, has a resolution equivalent to conodont zones for the Middle and Upper Devonian.

The initiation and diversification of vascular plants during the Devonian provides

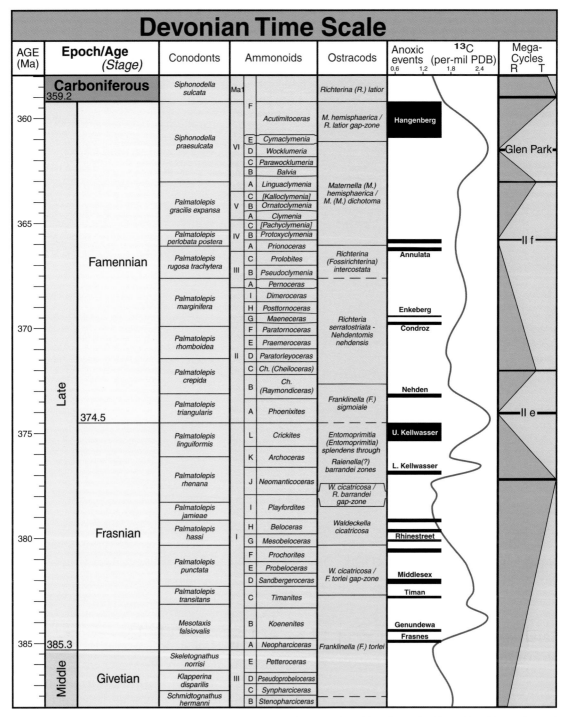

Figure 7.4. Numerical ages of epoch/series and age/stage boundaries of the Devonian with major marine biostratigraphic zonations and significant "Event" levels (e.g., widespread anoxic facies or important outcrops), principle eustatic trends, and ^{13}C isotopes. ["Age" is the term for the time equivalent of the rock-record "stage".] The conodont and ostracod scales are from Melchin *et al.* (GTS04) and the ammonite scale is from Becker and House (2000). The Mega cycles are from Johnson *et al.* (1985), in which *T* = transgression (rising sea level), *R* = regression (falling sea level). The ^{13}C isotope curve is from Buggisch and Joachimski (2006; their Europe composite).

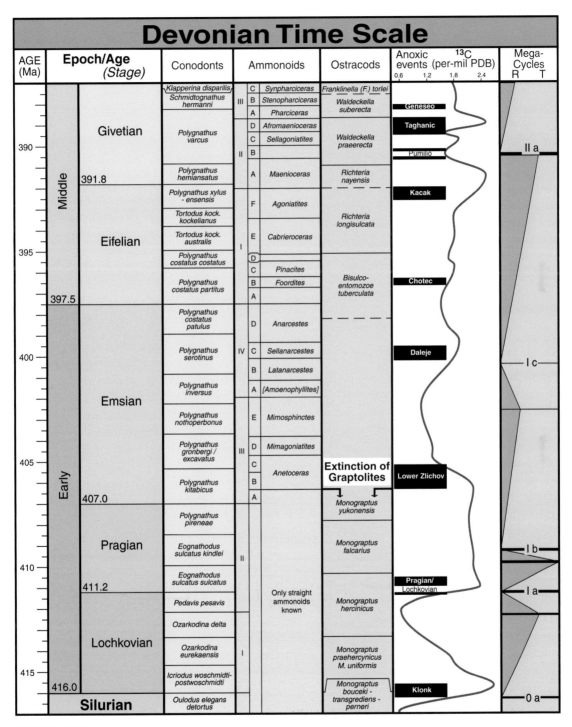

Figure 7.4. (cont.)

a miospore zonation for correlating marine and terrestrial strata. Fish evolved rapidly during the Devonian, and their jaws and armored remains provide a broad zonation.

Both marine and continental fauna and flora have a wide distribution, and there was little impact of biogeographic provinces. As a result, interbasinal correlations have been achieved with a relatively high degree of precision compared to the following Carboniferous through Triassic.

Anoxic events and carbon-isotope curve

Throughout the Devonian there were periods of widespread hypoxic or anoxic sedimentation (that is, sedimentary events indicated that little free oxygen or no oxygen at all was dissolved in Devonian seas). Some of these are known to be periods of significant extinction, and all are associated with some faunal anomaly in marine strata. Some are associated with very wide distribution of certain taxa, such as the *Monograptus uniformis*, *Pinacites jugleri* and *Platyclymenia annulata* events (Klonk, Chotec, and Annulata events, respectively). The Lower Zlichov Event is associated with the extinction of the graptolites and the appearance of the coiled goniatite ammonoids. Three anoxic events are very significant extinction episodes:

(1) The Taghanic Event, which formerly was used to draw the boundary between the Middle and Upper Devonian, was a marked period of extinction for goniatite ammonoids, corals and brachiopods.

(2) The Kellwasser Event at the Frasnian–Famennian boundary saw the extinction of the beloceratid and manticoceratid goniatite ammonoid groups, many conodont species, most colonial corals, several groups of trilobites, and the atrypid and pentamerid brachiopods.

(3) The Hangenberg Event at the end of the Famennian Stage corresponds to the extinction of phacopid trilobites, several groups of goniatite ammonoids, and the unusual Late Devonian coiled clymeniid ammonoids.

Most of the major anoxic events (Klonk, base Pragian, Taghanic, Lower and Upper Kellwasser, Hangenberg, etc.) are associated with rises in carbon-13, presumably due to increased burial of organic carbon-12. These carbon-isotope excursions are a powerful means for global correlation.

The increased frequency of anoxic events and a general shift toward carbon-13 enrichment during the Middle and Late Devonian may be the result of increased burial rate of organics through the combined effect of the evolution of tree-sized plants, more effective soil retention and weathering, large delta deposits and enhanced release of nutrients into the marginal seas.

Numerical time scale (GTS04 and future developments)

Conodont zones are the primary standard for global correlation of GSSPs and intercorrelation of most other marine and geochemical events. Michael House, a former Devonian specialist at Cambridge, had prepared a schematic estimate of relative durations of the standardized detailed conodont zones and subzones within each stage. The numerical time scale in GTS04 was derived by adjusting this proportional zonal scheme to fit the available array of radiometric ages. In turn, the calculated numerical ages for each conodont zone enabled interpolation of ages for the stage boundaries and other stratigraphic events.

Kaufmann (2006) used a modified version of this procedure after first estimating the relative durations of conodont zones from their relative thicknesses in selected stratigraphic sections or in stage-level graphic correlations of selected sections. In some cases, such as in the upper Emsian and upper Famennian, these relative durations are quite uncertain, since there is no support for zonal durations from graphic correlation of comparable sedimentary sections, tuned sedimentary cycles or subzonal duration constraints (Th. Becker and B. Kaufmann, pers. comm., 2006/2007). The major difference between the Devonian scales in GTS04 and in Kaufman (2006) is that base Eifelian is nearly 6 myr younger in the latter, and consequently the Emsian is much longer than in GTS04. Contributing to this long duration are the 392.2 ± 1.5 Ma age used for the Wetteldorf

"Hercules I" zircon tips, plus the interpreted very long duration for the "serotinus" conodont zone of the upper Emsian. Hydrogen fluoride leaching of the zircons without annealing may have changed the U–Pb content (F. Corfu, pers. comm., 2007), and the real age may be between 400 and 390 Ma, with an uncertainty that covers 10 myr or less.

We emphasize the need for acquisition of additional radiometric ages within each Devonian stage, using proper error analysis, and for the compilation of a global composite standard of conodont zoned strata that strives to remove possible distortions due to sea-level and other regional sedimentation-rate influences.

Acknowledgements

For further details/information, we recommend "The Devonian Period" by M. House and F. M. Gradstein (in *A Geologic Time Scale 2004*) and "Devonian Period" by Michael R. House in *Encyclopedia Britannica*. Portions of the background material are from documents of the Devonian Subcommission.

Further reading

Becker, R. T., and House, M. R., 2000. Devonian ammonoid zones and their correlation with established series and stage boundaries. *Courier Forschungsinstitut Senckenberg* **220**: 113–151.

Buggisch, W., and Joachimski, M. M., 2006. Carbon isotope stratigraphy of the Devonian of Central and Southern Europe.

Palaeogeography, Palaeoclimatology, Palaeoecology **240**: 68–88.

Bultynck, P. (ed.), 2000. *Recognition of Devonian Series and Stage Boundaries in Geological Areas, Courier Forschungsinstitut Senckenberg*, vol. 225. Frankfurt am Main: Forschungsinstitut und Naturmuseum Senckenberg.

House, M. R., 2002, reprinted 2007. Devonian Period. *Encyclopedia Britannica*. Available on-line at *www.britannica.com*

Johnson, J. G., Klapper, G., and Sandberg, C. A., 1985. Devonian eustatic fluctuations in Euramerica. *Geological Society of America Bulletin* **96**: 567–587.

Kaufmann, B., 2006. Calibrating the Devonian time scale: a synthesis of U–Pb ID-TIMS ages and conodont stratigraphy. *Earth-Science Reviews*, **76**: 175–190.

McGhee, G. R., Jr., 1996. *The Late Devonian Mass Extinction: The Frasnian/Famennian Crisis*. New York: Columbia University Press.

Menning, M., Alekseev, A. S., Chuvashov, B. I., Davydov, V. I., Devuyst, F.-X., Forke, H. C., Grunt, T. A., Hance, L., Heckel, P. H., Izokh, N. G., Jin, Y.-G., Jones, P. J., Kotlyar, G. V., Kozur, H. W., Nemyrovska, T. I., Schneider, J. W., Wang, X.-D., Weddige, K.,

Weyer, D., and Work, D. M., 2006. Global time scale and regional stratigraphic reference scales of Central and West Europe, East Europe, Tethys, South China, and North America as used in the Devonian–Carboniferous–Permian Correlation Chart 2003 (DCP 2003). *Palaeogeography, Palaeoclimatology, Palaeoecology*, **240**: 318–372.

Ziegler, P. A. 1988. Laurussia: the Old Red Continent. In: *Devonian of the World*, eds. N. J. McMillan, A. F. Embry, and D. J. Glass. *Canadian Society of Petroleum Geologists Memoir* **14**(1): 15–48.

Selected on-line references

Devonian Subcommission – *www.unica.it/sds*

Devonian Times website – *www.devoniantimes. org/* – [a fun one that was maintained by Dennis Murphy and received Science & Technology web award from *Scientific American* in 2005; but is no longer updated regularly]

We recommend the extensive Devonian webpages and links at *Palaeos*, Smithsonian Institution, University of California Museum of Paleontology, and *Wikipedia*. See URL details at end of Chapter 1.

8 Carboniferous Period

Philip H. Heckel

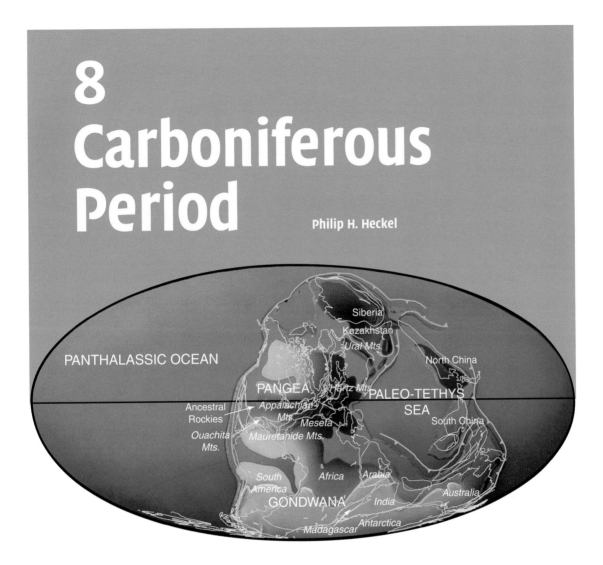

Figure 8.1. Geographic distribution of the continents during the Carboniferous Period (306 Ma). The paleogeographic map was provided by Christopher Scotese.

History and base of Carboniferous

The first use of the name Carboniferous for the rock succession to which it now applies is attributed to William Conybeare and William Phillips in 1822 for coal-bearing strata in their outline of the geology of England and Wales. It was the first system to receive a name that is still in official use today.

The Devonian–Carboniferous boundary falls within an interval of global regression following the major extinction associated with the Hangenberg anoxic event. It had been designated as the base of the *Gattendorfia* ammonoid Zone in 1937. The IUGS-ratified boundary GSSP established in 1991 at La Serre, France, at the first appearance of the conodont *Siphonodella sulcata*, is only slightly older than the classic ammonoid boundary. Recent detailed study has revealed problems with this GSSP, and a new Devonian–Carboniferous Boundary Task Group is being selected to suggest modifications.

Figure 8.2. The GSSP marking the base of the Carboniferous System and its lowermost Tournaisian Stage, at La Serre, France. The section is vertically dipping. The photograph was provided by R. Feist.

International subdivisions of Carboniferous

The Carboniferous has undergone an extremely varied history of subdivision in different regions. The basic subdivision used predominantly in eastern Europe and Asia was tripartite, and the units were referred to by the positional terms Lower, Middle, and Upper, with the rank of series. Elsewhere the subdivision was bipartite, as in western Europe, where the names Lower and Upper (later Dinantian and Silesian) were applied with the rank of subsystem. The subdivision was exclusively bipartite in North America, where the names Mississippian and Pennsylvanian were applied in the United States, with the rank of system throughout the twentieth century. The Eurasian subsystem and series terms were used with widely differing boundaries in different areas, whereas the American names were applied consistently, as they had a typically disconformable contact between them. Thus, this contact in an area where it was essentially conformable was selected as the Mid-Carboniferous boundary, and the terms Mississippian and Pennsylvanian

were voted to be the official global names of the two subsystems in 1999–2000. The extent of a possible stratigraphic gap between these subsystems in Eurasian stratigraphy remains uncertain.

Each of the subsystems became subdivided in 2003–2004 into three series with the positional names Lower, Middle, and Upper Mississippian, and Lower, Middle, and Upper Pennsylvanian. The component global stages have maintained the names that had long been used in eastern Europe and parts of Asia (Table 8.1).

Selected aspects of Carboniferous stratigraphy

Cyclic and sequence stratigraphy

Most Pennsylvanian and upper Mississippian strata are characterized by marine cyclothems that resulted from glacially controlled sea-level changes. In the shelf regions of midcontinent and eastern North America and eastern Europe, in certain basins at the proper elevations (e.g., Donets), and on carbonate banks in

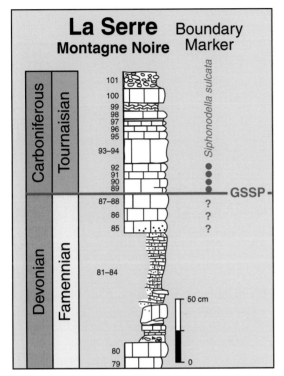

Figure 8.3. Stratigraphy of the basal Carboniferous GSSP in the section at La Serre, France, with the primary boundary markers. The GSSP is currently fixed at the base of Bed 89, which was based on the supposed first *Siphonodella sulcata* in a morphological transition from *S. praesulcata*. However, siphonodellids that cannot be distinguished morphologically from the supposed first *S. sulcata* have been found during resampling as low as in the upper part of Bed 84 and transitional specimens between *S. sulcata* and *S. duplicata* occur first in Bed 86. This suggests that the GSSP was positioned in the upper part of the *S. sulcata* Zone, a level that elsewhere is younger than the onset of the *Gattendorfia* ammonoid faunas that traditionally ("Heerlen decision": Jongmans and Gothan, 1937) and uncontestedly are regarded as Carboniferous. A precise correlation of the GSSP at the base of Bed 89 into other sections has become impossible since there is no alternative biostratigraphical marker, no distinctive morphotype within *S. sulcata*, and since all conodonts have possibly been subject to reworking and re-deposition within Bed 89.

tectonically disturbed regions (e.g., Spain, China, southwestern USA), these highstand cyclothems are separated by lowstand disconformities. Therefore, such cyclothems are stratigraphic sequences that can be "digitally" correlated by a combination of biostratigraphy and relative position in the succession. Cyclothems of major, intermediate, and minor scale are interpreted as resulting from the interaction of the periodically varying orbital parameters of the Earth's orbit, which caused fluctuations in ice caps on Gondwana. Major cyclothem groupings of 400 kyr (long eccentricity), each containing one major cyclothem, encompass oscillations of 100 kyr (short eccentricity), 30 kyr (Pennsylvanian obliquity), and 20 kyr (precession) (Strasser *et al.*, 2006).

Therefore, this cyclic stratigraphy is potentially the best method for both high-resolution correlation within and among basins and for scaling the durations of mid to late Carboniferous biostratigraphic zones. The greatest highstands are represented by the major cyclothems, which can be correlated globally throughout the ancient tropical belt (Heckel *et al.*, 2007). The greatest lowstands, which occur episodically between certain of the major cyclothem groupings, should be recognized by more widespread glacial deposits across much of the Gondwana region (Heckel, 2008).

Biostratigraphy

Although ammonoid cephalopods were used early in Europe to subdivide and zone the Carboniferous, their general restriction to deeper-water facies inhibited their use across the broad shelves of eastern Europe and North

Table 8.1 GSSPs of Carboniferous stages, with location and primary correlation criteria (status in 2008)

Stage	GSSP location	Latitude, longitude	Boundary level	Correlation events	Reference
Gzhelian	Candidates are in southern Urals and Nashui, S China			Conodont FAD of *diognathodus simulator* (s.str.). Close to ammonoid FAD of *Shumardites*	
Kasimovian	Candidates are in southern Urals, southwest USA, and Nashui, S China			Fusulinid FAD of *Protriticites*, which is near ammonoid FAD of *Eothalossoceras*. Alternative (higher) base is fusulinid FAD of *Montiparus montiparus*, which is near conodont FAD of *Idiognathodus sagittalis*. *Age given here is the lower version; the higher one is about 1 myr younger*	
Moscovian	Candidates are in southern Urals and Nashui, S China			Either conodont FAD of *Idiognathoides postsulcatus* or *Declinognathodus donetzianus*	
Bashkirian	Arrow Canyon, Nevada	36° 44′ 00″ N 114° 46′ 40″ W[b]	82.9 m above the top of the Battleship Formation in the lower Bird Spring Formation	Conodont FAD *Declinograptus noduliferous*	*Episodes* **22** (4), 1999
Serpukhovian	Candidates are Verkhnyaya Kardailovka, Urals, Nashui, S China			Conodont FAD of *Lochriea ziegleri*	
Visean	Pengchong, S China	24° 26′ N 109° 27′ E	Base of Bed 83 in the Pengchong Section	Foraminifer FAD of *Eoparastaffella simplex*	
Tournaisian *(base Carboniferous)*	La Serre, France	43° 33′ 19.9″ N 3° 21′ 26.3″ E[a]	Base of Bed 89 in Trench E′ at La Serre *(but FAD now known to be at base of Bed 85)*	Conodont FAD *Siphonodella sulcata* IMPRECISE (GSSP discovered in 2006 to have biostratigraphic problems, and can not be correlated with precision)	*Episodes* **14** (4), 1991; *Kölner Forum Geol. Paläont.* **15**, 2006

a. According to Google Earth.
b. Derived from map.
Source: Details on each GSSP are available at *www.stratigraphy.org* and in the *Episodes* publications.

America. As a result, microscopic foraminifers, especially the larger fusulinids in the Pennsylvanian, became used in those regions to subdivide and correlate the Carboniferous. Unfortunately their shallow-benthic habitat rendered foraminifers provincial throughout the entire Carboniferous, inhibiting their use in global correlation. Therefore, pelagic conodont microfossils have been used more recently for global correlation of the Carboniferous. During the Early Mississippian, and particularly in the Pennsylvanian when glacial–eustatic fluctuations in sea level produced cyclothems with interglacial highstand deposits, certain conodont species gained global distribution on the broad shelves of midcontinent North America and eastern Europe.

Plant remains and spores in coal-bearing strata are used in terrestrial successions. Where such deposits interfinger with the marine succession in eastern North America and western Europe, these plant fossils allow correlation between the marine and terrestrial realms.

Stable-isotope stratigraphy

Stable-isotope stratigraphy uses global variations in the stable isotope ratios of carbon, oxygen and strontium to correlate regions. This is a valuable method for accurately correlating the strongly provincial cold-climate south-polar Gondwana region (India, Africa, Australia, Antarctica, southern South America) and the equally provincial north-polar region (e.g., Angara region of northeastern Asia) with the diversely fossiliferous pantropical regions (North America and central Eurasia) where the biostratigraphic zonations have been developed.

A major positive excursion in carbon-13 in the Early Mississippian is one of the largest (up to +7 per mil) in the Phanerozoic. A reported simultaneous drop in oxygen-18 values suggests that this late Tournaisian episode of enhanced carbon burial was accompanied by global cooling. The late Mississippian and Mississippian–Pennsylvanian boundary intervals appear to have several negative carbon-isotope excursions.

Numerical time scale (GTS04 and future developments)

Assignment of numerical ages to Carboniferous biostratigraphic events and stage boundaries in *A Geologic Time Scale 2004* utilized a composite standard from graphical correlation of different sections, especially those of eastern Europe and southern Urals. The composite standard incorporated conodont, ammonoid, and benthic foraminifer events. The primary scaling for the Mississippian interval was mainly constrained by benthic foraminifers, whereas the Pennsylvanian was constrained mainly by conodonts. This composite standard was fit to an array of selected radiometric ages (both Ar–Ar and U–Pb) using a cubic spline.

In the years since the compilation of GTS04, U–Pb ages from Pennsylvanian strata and preliminary cycle-stratigraphy interpretations have generally supported the GTS04 age estimates for stage boundaries, but indicate the need for adjusting the relative durations of component conodont zones. In addition, the working definitions of at least two of the boundaries have been revised.

Base of new definition of Serpukhovian is 328.39 Ma

In GTS04, the base of the Serpukhovian Stage had been assigned an age of 326 Ma based on a tentative working definition of the first appearance of the conodont *Lochriea cruciformis*. The boundary working group decided that the first evolutionary appearance of the conodont *L. ziegleri* in the lineage *L. nodosa – L. ziegleri* presents the best potential for an international definition of the boundary (Appendix B in Carboniferous Subcommission

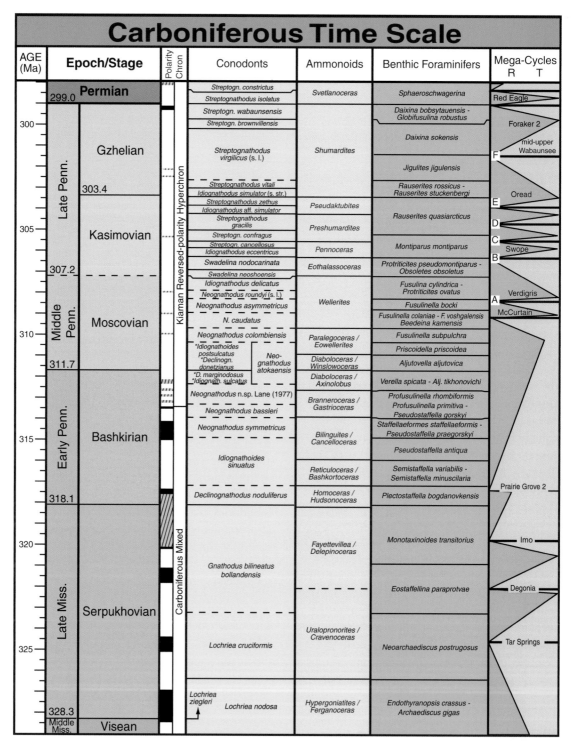

Figure 8.4. Numerical ages of epoch/series and age/stage boundaries of the Carboniferous with major marine biostratigraphic zonations and principle eustatic trends. ["Age" is the term for the time equivalent of the rock-record "stage".] Scaling is based on Russian foraminifer zones from GTS04. For the Pennsylvanian Subsystem, the conodont zonation is updated slightly from Barrick *et al.* (2004) for North America, the ammonoid zonation is updated from Boardman *et al.* (1994) for North America, and the foraminifer zonation is from Koren' [ed.], (2006) for Russia. For the Mississippian Subsystem, the conodont, ammonoid, and foraminifer zonation all are from Koren' [ed.], (2006) for Russia. The major sea-level sequences for the Lower and middle part of the Carboniferous are from Ross and Ross (1987, 1988). The Middle and Upper Pennsylvanian sea-level mega-cycles are derived from charts in Heckel (2008) (see details in Figure 8.5), for which the capital letters on the Moscovian through Gzhelian succession denote the greatest regressions when glacial deposits should be more widespread in the Gondwana region. Correlations are updated from Menning *et al.* (2006) and Heckel *et al.* (2007). Possible extent, if any, of a hiatus spanning the Mississippian–Pennsylvanian boundary interval in Eurasia is controversial.

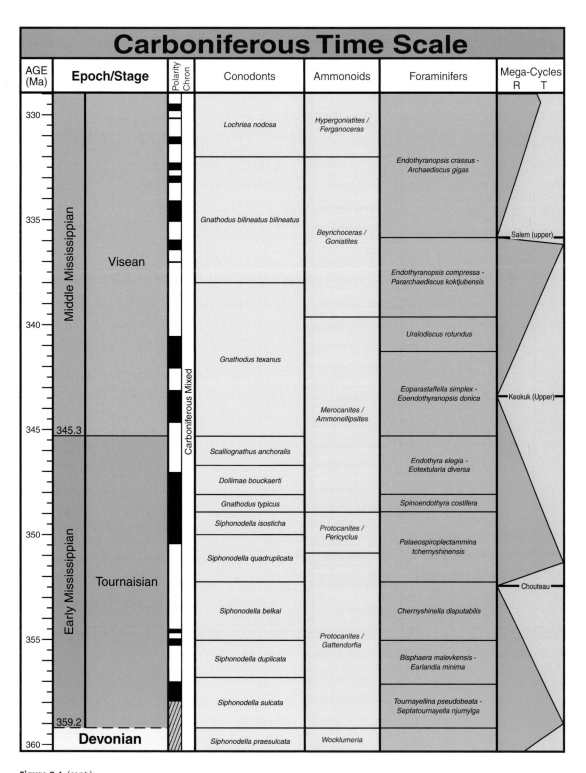

Figure 8.4. (cont.)

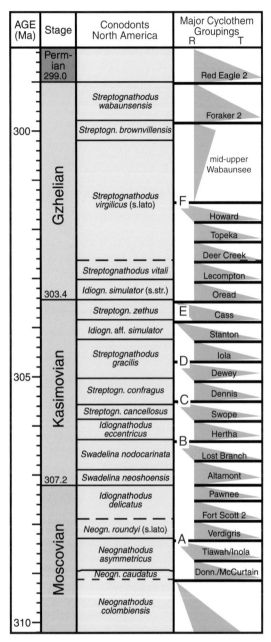

AGE (Ma)	Stage	Conodonts North America	Major Cyclothem Groupings R T
	Perm- ian 299.0		Red Eagle 2
		Streptognathodus wabaunsensis	Foraker 2
300		*Streptogn. brownvillensis*	
	Gzhelian		mid-upper Wabaunsee
		Streptognathodus virgilicus (s.lato)	F Howard
			Topeka
			Deer Creek
		Streptognathodus vitali	Lecompton
303.4		*Idiogn. simulator* (s.str.)	Oread
		Streptogn. zethus	E Cass
		Idiogn. aff. *simulator*	Stanton
	Kasimovian	*Streptognathodus gracilis*	D Iola
305			Dewey
		Streptogn. confragus	Dennis
		Streptogn. cancellosus	C Swope
		Idiognathodus eccentricus	Hertha
		Swadelina nodocarinata	B Lost Branch
307.2		*Swadelina neoshoensis*	Altamont
		Idiognathodus delicatus	Pawnee
			Fort Scott 2
	Moscovian	*Neogn. roundyi* (s.lato)	Verdigris
		Neognathodus asymmetricus	A Tiawah/Inola
		Neogn. caudatus	Donn./McCurtain
		Neognathodus colombiensis	
310			

Figure 8.5. Succession of ~400-kyr cyclothem groupings with respective conodont zones from mid-Moscovian through Gzhelian stages in midcontinent North America. Each grouping is centered around a single major cyclothem (or two in rare cases) and includes several cyclothems of lesser scale, all of which reflect interaction of Earth's orbital parameters. Names on right side of chart refer to major transgressive cyclothems in the groupings, slightly modified from Heckel (2008). Main part of succession from Pawnee through Lecompton is correlated with Eurasia by Heckel *et al.* (2007).

portion of 2007 ICS Annual Report). This level is below the current base of the Serpukhovian as defined by its type section near the town of Serpukhov in the Moscow Basin.

Base of Kasimovian is revised to 307.2 Ma and base of Gzhelian is revised to 303.4 Ma

GTS04 had estimated ages of 306.5 Ma for base Kasimovian and 303.9 Ma for base Gzhelian based on graphic correlation of conodont zones. The new ages are derived from cyclothem-calibrated correlation horizons for the potential GSSP primary correlation markers – the base of the *Protriticites pseudomontiparus – Obsoletes obsoletus* foraminifer Zone (equals base of *Swadelina neoshoensis* conodont Zone) for the current base of the Kasimovian, and lowest occurrence of *Idiognathodus simulator* sensu stricto now defining the base of the Gzhelian. This potential base of the Kasimovian Stage coincides with the transgression at the base of the Farlington (lower part of the Altamont cyclothem grouping) cycle of North American Midcontinent (= Lower Suvorovo of Moscow Basin, and cycle N3 of Donets Basin). However, the definition of the global Kasimovian is still under discussion. The base of the Gzhelian Stage coincides with the transgression at the base of the Oread cycle (Upper Rusavkino of Moscow Basin; O6 of Donets Basin) (Heckel *et al.*, 2007). Using 405-kyr long-eccentricity cycles relative to the base-Permian GSSP within this cyclothem framework derives the timings of these groupings.

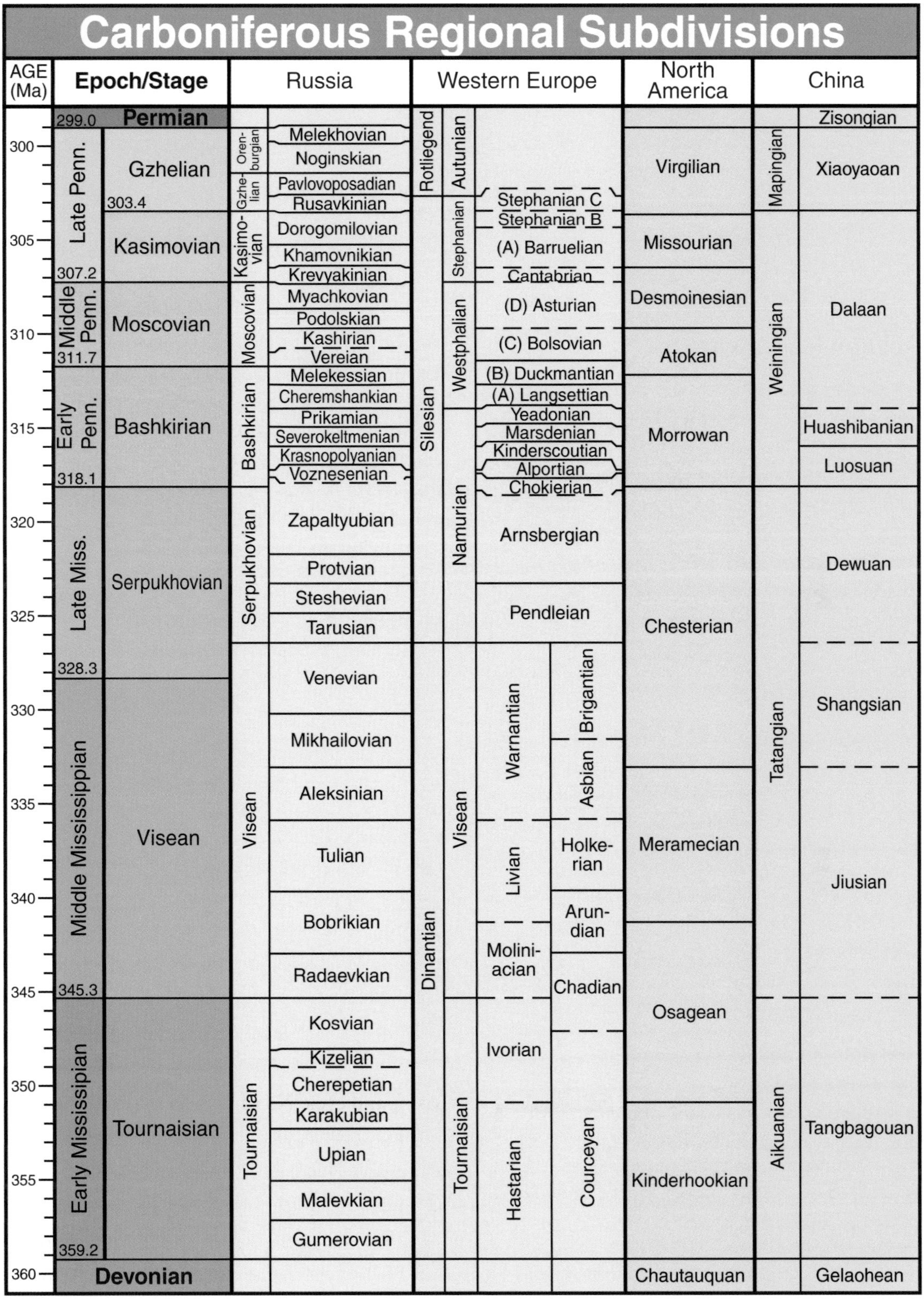

Carboniferous Regional Subdivisions

AGE (Ma)	Epoch/Stage	Russia	Western Europe	North America	China
299.0	**Permian**				Zisongian
	Late Penn. — Gzhelian (Orenburgian)	Melekhovian / Noginskian / Pavlovoposadian / Rusavkinian	Rotliegend — Autunian	Virgilian	Xiaoyaoan
303.4		(Gzhelian)			
	Kasimovian	Dorogomilovian / Khamovnikian / Krevyakinian	Stephanian C / Stephanian B / (A) Barruelian / Cantabrian	Missourian	
307.2		(Kasimovian)			
	Middle Penn. — Moscovian	Myachkovian / Podolskian / Kashirian / Vereian	Westphalian: (D) Asturian / (C) Bolsovian / (B) Duckmantian / (A) Langsettian	Desmoinesian / Atokan	Dalaan
311.7		(Moscovian)			
	Early Penn. — Bashkirian	Melekessian / Cheremshankian / Prikamian / Severokeltmenian / Krasnopolyanian / Voznesenian	Silesian / Namurian: Yeadonian / Marsdenian / Kinderscoutian / Alportian / Chokierian	Morrowan	Huashibanian
315		(Bashkirian)			Luosuan
318.1					
320	Late Miss. — Serpukhovian	Zapaltyubian / Protvian / Steshevian / Tarusian	Arnsbergian / Pendleian	Chesterian	Dewuan
328.3		(Serpukhovian)			
330	Middle Mississippian — Visean	Venevian / Mikhailovian / Aleksinian / Tulian / Bobrikian / Radaevkian	Dinantian — Visean: Warnantian / Brigantian / Asbian / Livian / Holkerian / Arundian / Moliniacian / Chadian	Meramecian / Osagean	Shangsian / Jiusian
345.3		(Visean)			
345	Early Mississippian — Tournaisian	Kosvian / Kizelian / Cherepetian / Karakubian / Upian / Malevkian / Gumerovian	Tournaisian: Ivorian / Hastarian / Courceyan	Kinderhookian	Tangbagouan
359.2		(Tournaisian)			
360	**Devonian**			Chautauquan	Gelaohean

China column subdivisions: Mapingian, Weiningian, Tatangian, Aikuanian.

Figure 8.6. Correlation of the international subdivisions of the Carboniferous System with selected regional stage and substage nomenclatures. Global Stages (with Russian names), western European stages and substages and North American stages are modified from Heckel and Clayton (2006), using certain relative stage and substage lengths from Menning *et al.* (2006) and from correlation with major cyclothem groupings. Russian substages are from GTS04. China is from Menning *et al.* (2006), except for Pennsylvanian, which is from Zhang and Zhou (2007).

Acknowledgements

For further details/information on the numerical time scale, we recommend "The Carboniferous Period" by V. Davydov, B. R. Wardlaw, and F. M. Gradstein (in *A Geologic Time Scale 2004*). Portions of the background material are from documents of the Carboniferous Subcommission. Peter Jones (Canberra, Australia) contributed to revision of the graphics.

Further reading

Barrick, J. E., Lambert, L. L., Heckel, P. H., and Boardman, D. R., 2004. Pennsylvanian conodont zonation for Midcontinent North America. *Revista Española de Micropaleontologia*, **36**: 231–250.

Boardman, D. R., Work, D. M., Mapes, R. H., and Barrick, J. E., 1994. Biostratigraphy of Middle and Late Pennsylvanian (Desmoinesian–Virgilian) ammonoids. *Kansas Geological Survey Bulletin*, **232**: 1–121.

Heckel, P. H., 2008. Pennsylvanian cyclothems in Midcontinent North America as far-field effects of waxing and waning of Gondwana ice sheets. In: *Resolving the Late Paleozoic Ice Age in Time and Space*, Geological Society of America Special Paper no. 441: 275–289.

Heckel, P. H., and Clayton, G., 2006. The Carboniferous System: use of the new official names for the subsystems, series, and stages. *Geologica Acta*, **4**: 403–407.

Heckel, P. H., Alekseev, A. S., Barrick, J. E., Boardman, D. R., Goreva, N. V., Nemyrovska, T. I., Ueno, K., Villa, E., and Work, D. M., 2007. Cyclothem ["digital"] correlation and biostratigraphy across the global Moscovian–Kasimovian–Gzhelian stage boundary interval (Middle–Upper Pennsylvanian) in North America and eastern Europe. *Geology*, **35**: 607–610.

Jongmans, W. J., and Gothan, W., 1937. Betrachtungen fiber die Ergebnisse des zweiten Kongresses fiir Karbonstratigraphie. *Deuxième Congrès International de Stratigraphie et de Géologie du Carbonifère*, Herleen, 1935, *Compte Rendu*, **1**: 1–40. [Heerlen decision.]

Koren', T. N. (ed.), 2006. *Biozonal stratigraphy of Phanerozoic in Russia*. St. Petersburg: VSEGEI-Press. (In Russian.)

Menning, M., Alekseev, A. S., Chuvashov, B. I., Davydov, V. I., Devuyst, F.-X., Forke, H. C., Grunt, T. A., Hance, L., Heckel, P. H., Izokh, N. G., Jin, Y.-G., Jones, P. J., Kotlyar, G. V., Kozur, H. W., Nemyrovska, T. I., Schneider, J. W., Wang, X.-D., Weddige, K., Weyer, D., and Work, D. M., 2006. Global time scale and regional stratigraphic reference scales of Central and West Europe, East Europe, Tethys, South China, and North America as used in the Devonian–Carboniferous–Permian Correlation Chart 2003 (DCP 2003). *Palaeogeography, Palaeoclimatology, Palaeoecology*, **240**: 318–372.

Ross, C. A., and Ross, J. R. P., 1987. Late Paleozoic sea levels and depositional sequences. In: *Timing and Depositional History of Eustatic*

Sequences: Constraints on Seismic Stratigraphy, eds. C. A. Ross and D. Haman. *Special Publications of the Cushman Foundation for Foraminiferal Research*, **24**: 137–149.

Ross, C. A., and Ross, J. R. P., 1988. Late Paleozoic transgressive–regressive deposition. In: *Sea-Level Changes: An Integrated Approach*, eds. C. K. Wilgus, B. S. Hastings, C. A. Ross, H. Posamentier, J. Van Wagoner, and C. G. St. C. Kendall. *SEPM Special Publications*, **42**: 227–247.

Strasser, A., Hilgen, F. J., and Heckel, P. H., 2006. Cyclostratigraphy: concepts, definitions, and applications. *Newsletters in Stratigraphy*, **42**: 75–114.

Zhang, L., and Zhou, J., 2007. Chronostratigraphic subdivision of the Upper Carboniferous in China. *Journal of Stratigraphy*, **31** (Suppl. 1): 105–107. [Abstracts of the 16th International Congress on the Carboniferous and Permian.]

Selected on-line references

Climate and the Carboniferous Period (by Monte Hieb, under Plant Fossils of West Virginia) – *mysite.verizon.net/mhieb/WVFossils/Carboniferous_climate.html*

We recommend the extensive Carboniferous webpages and links at *Palaeos*, Smithsonian Institution, University of California Museum of Paleontology, and *Wikipedia*. See URL details at end of Chapter 1.

Author

Philip H. Heckel, Department of Geoscience, University of Iowa, Iowa City, IA 52242, USA [Chair, Carboniferous Subcommission]

9
Permian Period

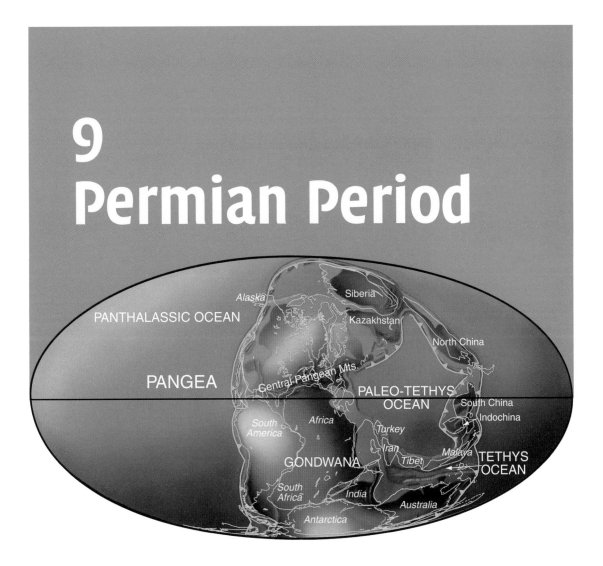

History and base of Permian

Roderick Murchison established the Permian in 1841, naming it after the ancient kingdom of Permia that formerly lay between the Volga River and the Ural Mountains in Russia. Initially, the base was placed at the initiation of evaporates, which is near the current base of the Kungurian Stage. The Permian System was later progressively extended downward to include the Artinskian (1889), then Sakmarian (1936), and finally Asselian (1954) stages.

The base of the Permian, as the base of the Asselian Stage, was placed near the appearance

Figure 9.1. Geographic distribution of the continents during the Permian Period (255 Ma). The paleogeographic map was provided by Christopher Scotese.

of inflated fusulinids of the *Schwagerina*-type group and three ammonoid families. However, the datum within the conodont lineage, which guided the placement of the GSSP within marginal marine facies in northern Kazakhstan, is slightly below these benthic foraminifer and ammonoid events. This conodont succession can be recognized with precision in many regions, including Texas and China.

Figure 9.2. The GSSP marking the base of the Permian System and its lowermost Asselian Stage at Aidaralash, Kazakhstan. The GSSP level is in the trench on the right, descending from the hilltop summit. Photo provided by Vladimir Davydov.

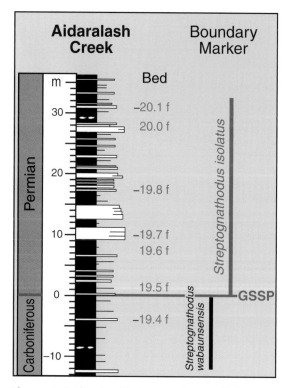

Figure 9.3. Stratigraphy of the base-Permian GSSP in the section at Aidaralash, Kazakhstan, with the primary boundary markers.

International subdivisions of Permian

The Permian naturally divides itself into three series, although the nomenclature for these series and subdivision into component stages varied greatly among regions. The Permian Subcommission established a global chronostratigraphic scheme in 1996 based on regional divisions and associated nomenclature of the lower Permian in the Urals, middle Permian in Texas of North America, and upper Permian of South China. All GSSPs were established in marine strata that preserved evolutionary lineages of conodonts. Most of the GSSPs are associated with other biostratigraphic, sequence-stratigraphic, or stable-isotopic events for inter-regional correlation.

The marine succession of lower Permian in the southern Urals (source of the "Cisuralian" Series name) of Russia and Kazakhstan has a rich biota with interstratified volcanic ashes that enable direct radiometric dating. The candidate suite of GSSPs in this region for the Sakmarian,

Table 9.1 GSSPs of Permian stages, with location and primary correlation criteria (status in 2008)

Stage	GSSP location	Latitude, longitude	Boundary level	Correlation events	Reference
Changhsingian	Meishan, Zhejiang Province, China	31° 4′ 55″ N 119° 42′ 22.9″ E	Base of Bed 4a-2, 88 cm above the base of the Changxing Limestone at the Meishan D Section	Conodont FAD *Clarkina wangi*	*Episodes* **29**(3), 2006
Wuchiapingian	Penglaitan, Guanxi Province, S China	23° 41′ 43″ N 109° 19′ 16″ E	Base of Bed 6 k in the Penglaitan Section	Conodont FAD *Clarkina postbitteri postbitteri*	*Episodes* **29**(4), 2006
Capitanian	Nipple Hill, SE Guadalupe Mountains, Texas, USA	31° 54′ 32.8″ N 104° 47′ 21.1″ W	4.5 m above the base of the outcrop section of the Pinery Limestone Member of the Bell Canyon Formation	Conodont FAD *Jinogondolella postserrata*	
Wordian	Guadalupe Pass, Texas, USA	31° 51′ 56.9″ N 104° 49′ 58.1″ W	7.6 m above the base of the Getaway Ledge outcrop section of the Getaway Limestone member of the Cherry Canyon Formation	Conodont FAD *Jinogondolella aserrata*	
Roadian	Stratotype Canyon, Texas, USA	31° 52′ 36.1″ N 104° 52′ 36.5″ W	42.7 m above the base of the Cutoff Formation	Conodont FAD *Jinogondolella nankingensis*	
Kungurian	Candidates are in southern Ural Mountains.			Near conodont FAD of *Neostreptognathus pnevi – N. exculptus*	
Artinskian	Candidates are in southern Ural Mountains			FAD of conodont *Sweetognathus whitei*	
Sakmarian	Candidate is at Kondurovsky, Orenburg Province, Russia			Near conodont FAD of *Sweetognathus merrelli*	
Asselian *(base Permian)*	Aidaralash Creek, Kazakhstan	50° 14′ 45″ N 57° 53′ 29″ E[a]	27 m above the base of Bed 19, Aidaralash Creek	Conodont FAD of isolated-nodular morphotype of *Streptognathodus* "*wabaunsensis*"	*Episodes* **21**(1), 1998

a. According to Google Earth.
Source: Details on each GSSP are available at *www.stratigraphy.org* and in the *Episodes* publications.

Artinskian and Kungurian stages will utilize evolutionary lineages of condonts as the primary correlation criteria. The conodont-rich strata have interbedded volcanic ashes that yield radiometric ages.

The middle Permian in the Guadalupe Mountains ("Guadalupian" Series) is an extensively studied, regional carbonate-rich succession that can be traced from shallow- to deep-water depositional settings. An evolutionary lineage of a single conodont genus (*Jinogondolella*) has guided the stratigraphic placement of GSSPs to delimit the Roadian, Wordian and Capitanian Stages within this succession.

A major regression and mass extinction terminates the Guadalupian Series, and continuous marine successions into the overlying Lopingian Series are rare. Indeed, the non-marine character of typical successions (e.g., the famous evaporite-rich Zechstein facies of central Europe)

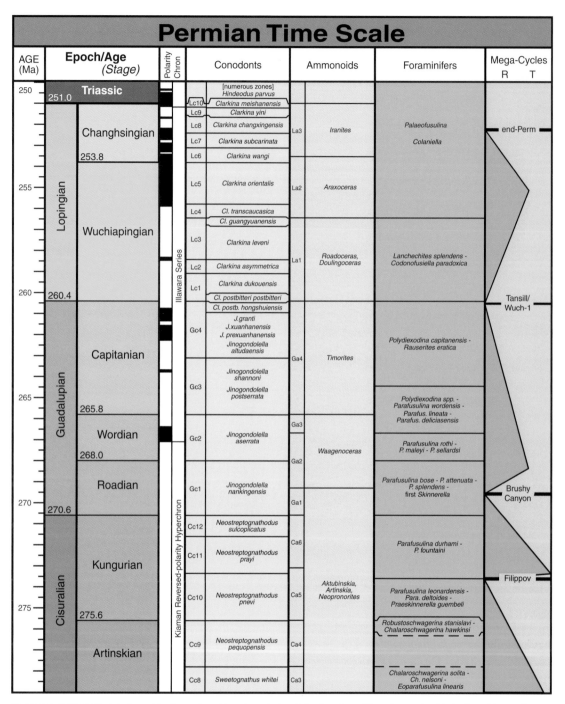

Figure 9.4. Numerical ages of epoch/series and age/stage boundaries of the Permian with major marine biostratigraphic zonations and principle eustatic trends. ["Age" is the term for the time equivalent of the rock-record "stage".] Biostratigraphic scales include conodont, ammonoid, and foraminifer zonations. The conodont scale is by Mei and Henderson (2001). The ammonite scale is from Kozur (2003), Davydov *et al.* (2004), and Henderson (2005). Foraminifers are from Davydov *et al.* (2004), Brenckle (in Lane and Brenckle, 2005, and pers. commun., Oct. 2006), and Ross and Ross (1988, 1995b). Magnetic polarity pattern is mainly from Steiner (2006). The Mega cycles are from Ross and Ross (1995a) and B. R. Wardlaw (unpubl. data).

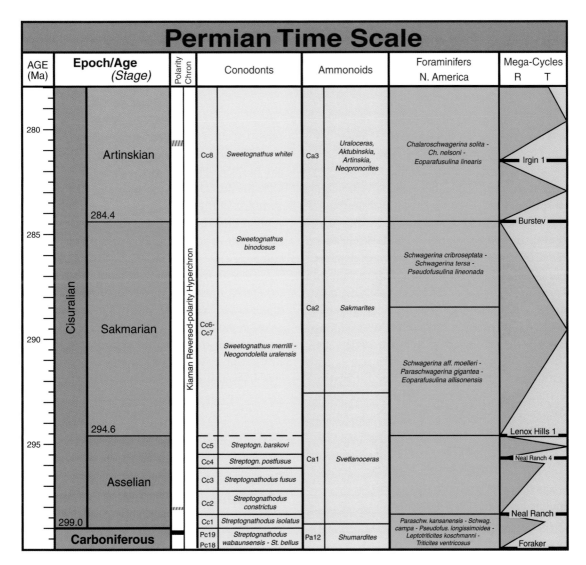

Figure 9.4. (cont.)

has led to long-standing debates on their correlation to the global standard and whether the strata encompass a significant part of the interval. The open-marine facies of the Lopingian Series of South China hosts the GSSPs for the Wuchiapingian and Changhsingian stages, and the selection of these horizons utilized evolutionary changes within the conodont genus *Clarkina*.

Selected aspects of Permian stratigraphy

Biostratigraphy

The trio of conodonts, benthic foraminifers (especially fusulinaceans) and ammonoids are the main biostratigraphic tools for correlation of marine facies. Different regional zonations are required for each of these groups. In the southern Urals, major sea-level lowstands coincide with significant fusulinacean extinctions, and highstands are preferential times of speciation.

Terrestrial biostratigraphy using plants (especially spores and pollen) and vertebrate remains is important for regional correlations, but correlation to the standard marine-based Permian stages has not yet been established in detail. Our knowledge of terrestrial tetrapod evolution is currently broken by "Olson's Gap" (approximately the entire Roadian Stage) between the well-known early Permian tetrapods from the western USA and the later Permian tetrapod records of Russia and South Africa.

Magnetic stratigraphy

An extended "Kiaman Reversed Superchron" spans the late Carboniferous into the middle Permian. The "Illawarra" interval of mixed polarity encompasses the upper Wordian through Changhsingian. A brief episode of normal polarity has been reported from near the Carboniferous–Permian boundary.

Sequence and stable-isotope stratigraphy

Fluctuations in the Gondwana glaciations produced frequent oscillations in sea levels during the Asselian and early Sakmarian. This was followed by relatively long intervals between major sequence boundaries through the middle Permian. An observed

extended interval of carbon-13 enrichment (the Kamura event) during the Capitanian has been interpreted as the product of high productivity in the tropical oceans with a resulting carbon-dioxide drawdown, cooling and the end-Guadalupian mass extinction of low-latitude warm-water fauna.

The terminology for major sea-level sequences shown in the Permian time-scale figure is derived from the chronostratigraphy of the regions hosting the reference sections; but this tentative sequence stratigraphy awaits confirmation. The relative lack of well-preserved marginal-marine successions in different regions inhibits the establishment of a robust global sequence stratigraphy.

Numerical time scale (GTS04 and future developments)

Assignment of numerical ages to Permian biostratigraphic events and stage boundaries in *A Geologic Time Scale 2004* utilized a composite standard from graphical correlation of different sections. The composite standard incorporated benthic foraminifer, conodont and ammonoid events; but the primary scaling was based on conodonts. This composite standard was fitted to an array of selected radiometric ages (both Ar–Ar and U–Pb) using a cubic spline.

The U–Pb ages from Upper Permian strata have been revised since the publication

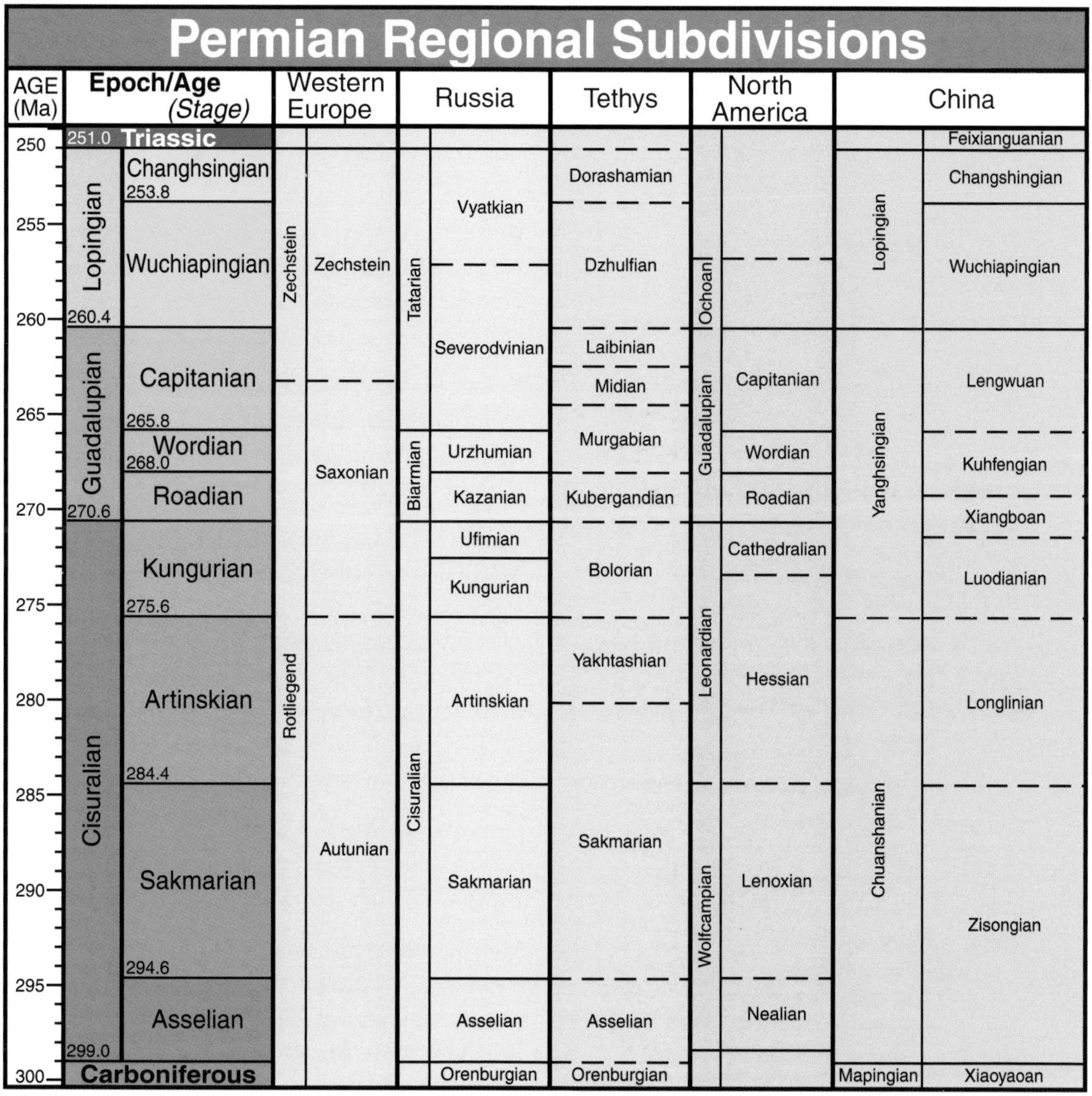

Permian Regional Subdivisions

AGE (Ma)	Epoch/Age (Stage)		Western Europe	Russia	Tethys	North America	China	
250		251.0 **Triassic**						Feixianguanian
	Lopingian	Changhsingian 253.8	Zechstein	Vyatkian	Dorashamian	Ochoan	Lopingian	Changshingian
255		Wuchiapingian	Zechstein	Tatarian	Dzhulfian			Wuchiapingian
260		260.4		Severodvinian	Laibinian			
	Guadalupian	Capitanian 265.8			Midian	Capitanian	Yanghsingian	Lengwuan
265		Wordian 268.0	Saxonian	Urzhumian	Murgabian	Wordian		Kuhfengian
270		Roadian 270.6		Kazanian	Kubergandian	Roadian		Xiangboan
		Kungurian		Ufimian	Bolorian	Cathedralian		Luodianian
275		275.6	Rotliegend	Kungurian				
280	Cisuralian	Artinskian		Artinskian	Yakhtashian	Hessian	Leonardian	Longlinian
285		284.4						
		Sakmarian	Autunian	Sakmarian	Sakmarian	Lenoxian	Chuanshanian	Zisongian
290						Wolfcampian		
295		294.6						
		Asselian 299.0		Asselian	Asselian	Nealian		
300		**Carboniferous**		Orenburgian	Orenburgian		Mapingian	Xiaoyaoan

Figure 9.5. Correlation of the international subdivisions of the Permian System with selected regional stage nomenclatures. Western Europe, Russia, Tethys are modified from Davydov (GTS04), and China is by Menning *et al.* (2006).

of GTS04 after an improvement in processing of zircons. In general, it now appears that the U–Pb-derived ages for events should be shifted about 1.5 myr older (e.g., the Permian–Triassic boundary is approximately 252.2 Ma, and the base of the Changhsingian is probably near 254.2 Ma. Several U–Pb radiometric ages directly tied to conodont zones of the Cisuralian Series are being obtained from the southern Urals, including strata near the proposed GSSPs for the lower Permian stages. The Carboniferous–Permian boundary has been reaffirmed as 299 Ma.

Acknowledgements

Charles Henderson (University of Calgary, chair of Permian Subcommission) contributed to this review. For further details/information, we recommend "The Permian Period" by B. R. Wardlaw, V. Davydov, and F. M. Gradstein (in *A Geologic Time Scale 2004*). Portions of the background material are from documents of the Permian Subcommission.

Further reading

Davydov, V., Wardlaw, B. R., and Gradstein, F. M., 2004. The Carboniferous Period, and The Permian Period. In: *Geologic Time Scale 2004*, eds. F. M. Gradstein, J. G. Ogg, and A. G. Smith. Cambridge: Cambridge University Press, pp. 222–270.

Henderson, C. M., 2005. International correlation of the marine Permian time scale. *Permophiles*, **46**: 6–9.

Isozaki, Y., Kawahata, H., and Ota, A., 2007. A unique carbon isotope record across the Guadalupian–Lopingian (Middle–Upper Permian) boundary in mid-oceanic paleo-atoll carbonates: the high-productivity "Kamura event" and its collapse in Panthalassa. *Global and Planetary Change*, **55**: 21–38.

Korte, C., Jasper, T., Kozur, H. W., and Veizer, J., 2005. δ18O and δ13C of Permian brachiopods: a record of seawater evolution and continental glaciation. *Palaeogeography, Palaeoclimatolology, Palaeoecology*, **224**: 333–351.

Korte, C., Brand, U., Dickins, J. M., Jones, P. J., Mertmann, D., and Veizer, J. (in press) Permian latitudinal sea-surface temperature gradients.

Kozur, H. W., 2003. Integrated Permian ammonoid, conodont, fusulinid, marine ostracod and radiolarian biostratigraphy. *Permophiles*, **42**: 24–33.

Lane, H. R., and Brenckle, P. L., 2005. Type Mississippian subdivisions and biostratigraphic succession In: *Stratigraphy and Biostratigraphy of the Mississippian Subsystem (Carboniferous system) in its Type Region, the Mississippi River valley of Illinois, Missouri, and Iowa*, ed. P. H. Heckel. *Illinois State Geological Survey Report*, **34**: 76–98.

Lucas, S. G., and Zeigler, K. E. (eds.), 2005. *The Nonmarine Permian. New Mexico Museum of Natural History and Science Bulletin* 30.

Mei, S., and Henderson, C. M., 2001. Evolution of Permian conodont provincialism and its significance in global correlation and paleoclimate implication. *Palaeogeography, Palaeoclimatology, Palaeoecology*, **170**: 237–260.

Menning, M., Alekseev, A. S., Chuvashov, B. I., Davydov, V. I., Devuyst, F.-X., Forke, H. C., Grunt, T. A., Hance, L., Heckel, P. H., Izokh, N. G., Jin, Y.-G., Jones, P. J., Kotlyar, G. V., Kozur, H. W., Nemyrovska, T. I., Schneider, J. W., Wang, X.-D., Weddige, K., Weyer, D., and Work, D. M., 2006. Global time scale and regional stratigraphic reference scales of

Central and West Europe, East Europe, Tethys, South China, and North America as used in the Devonian–Carboniferous–Permian Correlation Chart 2003 (DCP 2003). *Palaeogeography, Palaeoclimatology, Palaeoecology*, **240**: 318–372.

Ross, C. A., and Ross, J. R. P., 1988. Late Paleozoic transgressive–regressive deposition. In: *Sea-Level Changes: An Integrated Approach*, eds. C. K. Wilgus, B. S. Hastings, C. A. Ross, H. Posamentier, J. Van Wagoner, and C. G. St. C. Kendall. *SEPM Special Publications*, **42**: 227–247.

Ross, C. A., and Ross, J. R. P., 1995a. Permian sequence stratigraphy. In: *The Permian of Northern Pangea*, eds. P. A. Scholle, T. M. Peryt, and D. S. Ulmer-Scholle. Berlin: Springer, pp. 98–123.

Ross, C. A., and Ross, J. R. P., 1995b. Foraminiferal zonation of late Paleozoic depositional sequences. *Marine Micropaleontology*, **26**: 469–478.

Steiner, M. B., 2006. The magnetic polarity time scale across the Permian–Triassic boundary. In: *Non-Marine Permian Biostratigraphy and Biochronology*, eds. S. G. Lucas, G. Cassinis, and J. W. Schneider. *Geological Society of London Special Publications*, **265**: 15–38.

Selected on-line references

Permian Subcommission – *www.nigpas.ac.cn/permian/web/index.asp* – contains complete *Permophiles* issues

We recommend the extensive Permian webpages and links at *Palaeos*, Smithsonian Institution, University of California Museum of Paleontology, and *Wikipedia*. See URL details at end of Chapter 1.

10 Triassic Period

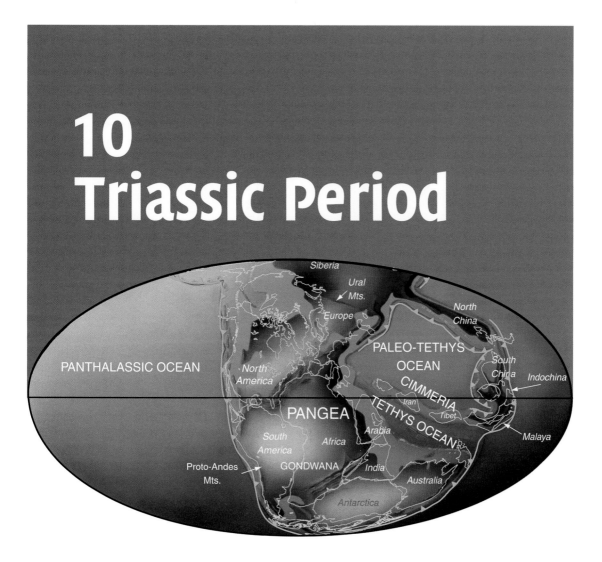

History and base of Triassic

The "Trias" of Friedrich August von Alberti (1834) united a trio of widespread formations in southern Germany (lower Buntsandstein, middle Muschelkalk, upper Keuper). However, the traditional stages of Anisian through Rhaetian were established in ammonoid-rich strata in the Austrian Alps; but these tectonic slices are generally unsuitable for establishing and delimiting formal stages.

The Paleozoic was terminated by a complex environmental and biologic catastrophe. The latest Permian through earliest Triassic events

Figure 10.1. Geographic distribution of the continents during the Triassic Period (237 Ma). The paleogeographic map was provided by Christopher Scotese.

include a pronounced negative excursion in carbon and strontium isotopes, the eruption of the massive Siberian Traps, widespread anoxia onto shelf environments and progressive disappearance of up to 80% of marine genera. The base of the Triassic has been defined as the initial stage of recovery from this end-Permian episode, and utilizes the first occurrence of the conodont *Hindeodus parvus* as the primary correlation marker from the GSSP section at Meishan, China.

Figure 10.2. The GSSP marking the base of the Triassic System and its lowermost Induan Stage at Meishan, China. The statue is topped by a sculpture of conodont *Hindeodus parvus*, which is the primary global correlation marker. Stairs lead to a platform at the GSSP outcrop. This impressive GeoPark also contains a modern museum of Earth's history.

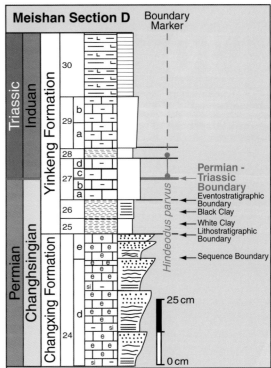

Figure 10.3. Stratigraphy of the base-Triassic GSSP in the section at Meishan, China, with the primary boundary markers.

This Meishan section also hosts the GSSP for the underlying Changhsingian Stage of uppermost Permian, and is now within a special GeoPark that includes a new museum of Earth's history.

International subdivisions of Triassic

Even though the nomenclature for Triassic stages was established over a century ago and formally adopted in 1992 as the international standard, the identification of global definitions for these stages through precise GSSPs has been a challenge. The search has been the driver of extensive international collaboration and research, and the placement of GSSPs for all the Triassic stages is now near completion. The traditional ammonite-based definitions for these stages within Europe is not always suitable for inter-regional correlation, therefore the established or candidate GSSP placements are enhanced by conodont, bivalve, paleomagnetic, isotopic and other secondary correlation horizons. The current candidates for Triassic stage boundaries are listed in Table 10.1.

Selected aspects of Triassic stratigraphy

Biostratigraphy

Conodonts, which are the phosphatic jaw elements of an enigmatic lamprey-like vertebrate,

Table 10.1 GSSPs of Triassic stages, with location and primary correlation criteria (status in 2008)

Stage	GSSP location	Latitude, longitude	Boundary level	Correlation events	Reference
Rhaetian	Key sections in Austria, British Columbia, Canada, and Turkey			Near ammonite FAD of *Cochloceras*, conodonts *Misikella* spp. and *Epigondolella mosheri*, and radiolarian *Proparvicingula moniliformis*	
Norian	Candidates are Black Bear Ridge in British Columbia, Canada, and Pizzo Mondello, Sicily, Italy			Base of *Stikinoceras kerri* ammonoid zone and near FAD of *Metapolygnathus echinatus* within the *M. communisti* conodont zone	
Carnian	Prati di Stuores, Dolomites, Italy. Important reference sections in Spiti, India, and New Pass, Nevada, USA	46° 31′ 37″ N 11° 55′ 49″ E	Base of marly limestone bed SW4, 45 m from base of San Cassiano Formation	Near FADS of the ammonoid *Daxatina*, the conodont '*Paragondolella*' *polygnathiformis*, and *Halobia* bivalves. Just above base of S2n magnetic polarity zone and above the maximum flooding surface of Sequence Lad 3 of Hardenbol et al. (1998)	*Albertiana* **36**, 2007
Ladinian	Bagolino, Province of Brescia, Northern Italy	45° 49′ 09.5″ N 10° 28′ 15.5″ E	Base of a 15–20-cm thick limestone bed overlying a distinctive groove ("Chiesense groove") of limestone nodules in a shaly matrix, located about 5 m above the base of the Buchenstein Beds	Ammonite FAD of *Eoprotrachyceras curionii* (base of the *E. curionii* zone); conodont FAD of *Budurovignathus praehungaricus* is in the uppermost Anisian	*Episodes* **28**(4), 2005
Anisian	Candidate section at Desli Caira, Dobrogea, Romania; significant sections in Guizhou Province, China, and South Primorye, Russia	45° 04′ 27″ N 28° 48′ 08″ E	In Section B, the GSSP level will be either the FAD of conodont *Chiosella timorensis* at the base of Bed GR7 at ~7 m; OR the base of magnetozone MT1n at the 5.7 m level.	Conodont FAD of *Chiosella timorensis* or Magnetic – base of magnetic normal-polarity chronozone MT1n	*Albertiana* **36**, 2007
Olenekian	Candidate GSSP Mud (Muth) village, Spiti Valley, NW India	31° 57′ 55.5″ N 78° 01′ 28.5″ E	Base of Bed 13A–2, about 4.8 m up in Mikin Formation, Section M04 (~4000 m elevation)	Conodont FAD of *Neospathodus waageni*, just above base of *Rohillites rohilla* ammonite zone, and below lowest occurrence of *Flemingites* and *Euflemingites* ammonite genera. Within a prominent positive carbon-13 peak, and just above widely recognizable sequence boundary	*Albertiana* **36**, 2007
Induan (*base Triassic*)	Meishan, Zhejiang Province, China	31° 4′47.28″N 119° 42′ 20.90′ E	Base of Bed 27c in the Meishan Section	Conodont FAD *Hindeodus parvus*	*Episodes* **24**(2), 2001

Source: Details on each GSSP are available at *www.stratigraphy.org* and in the *Episodes* publications.

have become as important as ammonoids in the correlation of Triassic marine facies. Both of these groups display pronounced regional provincialism during most of the Triassic, therefore the exact correlation of the inter-regional scales remains uncertain. The distinctive bivalves of the *Daonella*, *Halobia* and *Monotis* genera are important for subdividing the upper Anisian through Ladinian, the Carnian through mid-Norian and the uppermost Triassic, respectively. Spores and pollen play a key role in correlating terrestrial and marine strata.

The first appearance of the "mammal-like" dicynodont *Lystrosaurus* reptile marks the base of the Triassic in terrestrial settings, and a succession of mammal-like dicynodont and cynodont taxa enable a broad subdivision of Lower and Middle Triassic continental facies.

A major biologic and climatic change during the late Carnian (~230 Ma in the revised Triassic scale) coincided with the emergence of the earliest dinosaurs on land and the earliest calcareous nannoplankton in the oceans.

Stable-isotope stratigraphy

Major excursions in carbon, sulfur and strontium isotopes in uppermost Permian and lower Triassic strata are well documented, but the detailed patterns through the rest of the Triassic await refinement. An abrupt drop in ammonoid and conodont diversity at the base of the Spathian substage is associated with a major positive excursion in carbon-13, and there are suggestions that similar biotic–isotopic excursions occur near the beginnings of the Smithian, Anisian and Carnian. The Carnian shift initiated a pronounced broad positive carbon-13 plateau that gradually declines through the Norian. The boundary interval between Triassic and Jurassic has two carbon-13 excursions that enable inter-regional correlation of events (see Jurassic chapter).

Cycle and magnetic stratigraphy

The climate of the Pangea megacontinent seemed particularly responsive to Milankovitch cycles, especially the precession–eccentricity components of these orbital-climate oscillations. Variations in clastic input into the early Triassic Buntsandstein basins of central Europe seem to be governed by the 100-kyr eccentricity cycle, and closed-basin lakes in the middle to late Triassic Newark basins of easternmost North America record enhanced precipitation–evaporation shifts of the full precession–eccentricity cycle suite. Both of these continental basins have yielded excellent records of geomagnetic polarity. The correlation of this astronomical-tuned terrestrial magnetostratigraphy to the polarity pattern obtained from fossiliferous marine limestone sections is the basis of estimating relative durations of ammonite and conodont zones through most of the Triassic. However, the comparison of the cycle-scaled terrestrial polarity signature to the unscaled marine magnetostratigraphy does not always provide a unique match, as explained below.

Numerical time scale (GTS04 and future developments)

A Geologic Time Scale 2004 used a database of radiometric dates published prior to 2003 for constraining the ages of base Triassic (251 Ma, based especially on Bowring *et al.*, 1998), of base Ladinian (237 Ma, based on 240 Ma ages from mid-upper Anisian by Pálfy *et al.*, 2003), and base Jurassic (199.6 Ma, based on Pálfy *et al.*, 2000). The scaling of Early Triassic used a composite standard of conodont events, and the scaling of late Ladinian through Norian used a pattern match of magnetostratigraphy of Tethyan marine sections to the cycle-scaled terrestrial Newark magnetic polarity chrons similar to the "Option #1 Long Carnian" correlation of Muttoni *et al.* (2004). Intervening intervals were scaled using equal-duration ammonite subzones.

Implications of revised definitions of Triassic stages using GTS04 time scale

(1) Base of Carnian adjusted to 228.7 Ma. The Triassic Subcommission has decided to lower the base of Carnian to include the base of *D. canadensis* ammonite zone (base of *Daxatina*) and the onset of *Halobia* bivalves, instead of the higher FAD of *Trachyceras* ammonites (which was used in GTS04).

(2) Base of Anisian adjusted to 245.9 Ma, and base of Olenekian adjusted to 249.5 Ma. The Anisian–Olenekian boundary working group is considering the profound conodont turnover including lowest occurrences of *Chiosella* (*Cs. gondolelloides*) to be the primary marker for the base Anisian. GTS04 had used an ammonite placement near the higher FAD of *Cs. timorensis*. The base-Olenekian primary marker will be the lowest occurrence of conodont *Neospathodus waageni* sensu lato, just above base of the *Rohillites rohilla* ammonite zone. This is slightly different from in GTS04 (FAD of *Hedenstroemia* or *Meekoceras gracilitatis* ammonites). The ages for these base-Olenekian and base-Anisian levels are derived from a cycle scaling relative to the base Triassic (using its GTS04 age of 251.0 Ma).

Potential major modification of Triassic time scale

During and since the publication of GTS04, several exciting advances have significantly revised the constraints on Triassic time scale. First, new procedures of obtaining U–Pb ages from zircons have shifted ages to older values (e.g., the same base-Triassic levels that constrained the U–Pb age of 251.0 ± 0.4 Ma are now suggesting an age of between 252.5 ± 0.2 (Mundil *et al.*, 2004) and 252.2 Ma

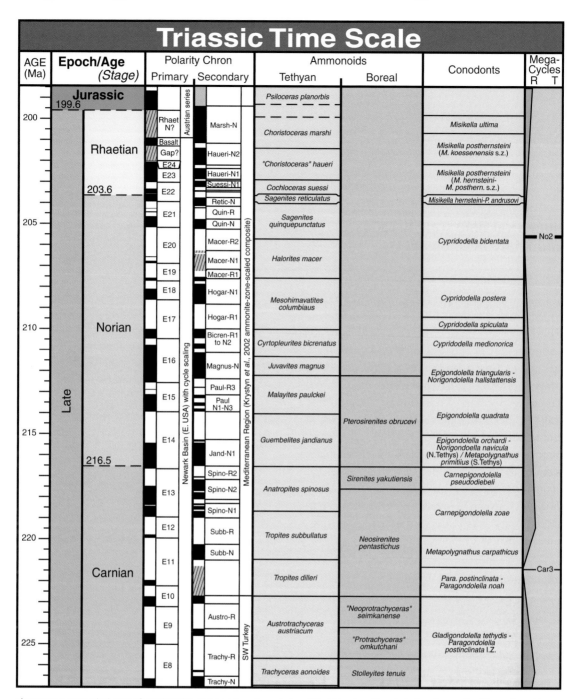

Figure 10.4. Numerical ages of epoch/series and age/stage boundaries of the Triassic with selected major marine biostratigraphic zonations and principle eustatic trends. ["Age" is the term for the time equivalent of the rock-record "stage".] The numerical time scale is from *A Geologic Time Scale 2004*; but see Figure 10.5 for a possible revision of these ages. Biostratigraphic scales include ammonoid and conodont zonations. The Tethyan ammonite scale is compiled from different sources, especially with revisions by Kozur (Kozur and Bachmann, 2005). The Tethyan conodont scale is by Kozur (2003, and pers. comm., 2006), with the Early Triassic portion modified after Orchard (2007) by Bob Nicoll; however, there is not universal agreement on the genera assignments for taxa nor on zonal divisions. The sea-level megacycles are from Hardenbol, J., Jacquin, T., Vail, P. R., *et al.* (SEPM charts. 1998). The magnetic polarity pattern is a composite; with Lower Triassic generalized from Szurlies (2004, 2007) and Hounslow (Hounslow *et al.*, 2007a, b), the Anisian through Norian is modified from Muttoni *et al.* (2004) and their correlations the orbital-scaled magnetostratigraphy from the Newark lake beds (Kent and Olsen, 1999), and the latest Norian and Rhaetian is modeled after Gallet *et al.* (2007). These scales have been undergoing rapid enhancements and revisions by the members of the Triassic Subcommission, and updates are published in *Albertiana*.

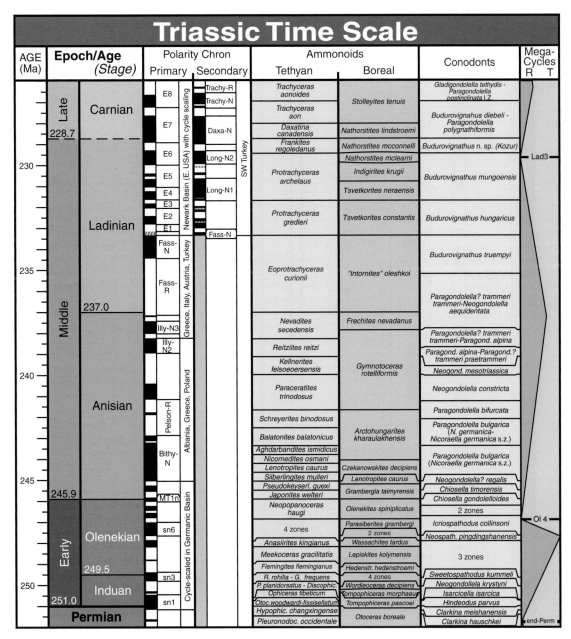

Triassic Time Scale

AGE (Ma)	Epoch/Age (Stage)	Polarity Chron		Ammonoids		Conodonts	Mega-Cycles
		Primary	Secondary	Tethyan	Boreal		R T
	Late — Carnian	E8	Trachy-R	Trachyceras aonoides	Stolleyites tenuis	Gladigondolella tethydis - Paragondolella postinclinata l.Z.	
			Trachy-N	Trachyceras aon		Budurovignathus diebeli - Paragondolella polygnathiformis	
228.7		E7	Daxa-N	Daxatina canadensis	Nathorstites lindstroemi		
230		E6	Long-N2	Frankites regoledanus	Nathorstites mcconnelli	Budurovignathus n. sp. (Kozur)	Lad3
		E5		Protrachyceras archelaus	Nathorstites mclearni	Budurovignathus mungoensis	
		E4	Long-N1		Indigirites krugii		
		E3			Tsvetkorites neraensis		
	Ladinian	E2		Protrachyceras gredleri	Tsvetkorites constantis	Budurovignathus hungaricus	
		E1	Fass-N				
		Fass-N				Budurovignathus truempyi	
235	237.0	Fass-R		Eoprotrachyceras curionii	"Intornites" oleshkoi		
						Paragondolella? trammeri trammeri-Neogondolella aequidentata	
		Illy-N3		Nevadites secedensis	Frechites nevadanus		
		Illy-N2		Reitziites reitzi		Paragondolella? trammeri trammeri-Paragond. alpina	
				Kellnerites felsoeoersensis	Gymnotoceras rotelliformis	Paragond. alpina-Paragond.? trammeri praetrammeri	
240	Anisian					Neogond. mesotriassica	
				Paraceratites trinodosus		Neogondolella constricta	
		Pelson-R		Schreyerites binodosus		Paragondolella bifurcata	
				Balatonites balatonicus	Arctohungarites kharaulakhensis	Paragondolella bulgarica (N. germanica-Nicoraella germanica s.z.)	
				Aghdarbandites ismidicus			
		Bithy-N		Nicomedites osmani		Paragondolella bulgarica (Nicoraella germanica s.z.)	
				Lenotropites caurus	Czekanowskites decipiens		
245				Silberlingites mulleri	Lenotropites caurus	Neogondolella? regalis	
	245.9	MT1n		Pseudokeyserl. guexi	Grambergia taimyrensis	Chiosella timorensis	
				Japonites welteri		Chiosella gondolelloides	
				Neopopanoceras haugi	Olenekites spiniplicatus	2 zones	
					Parasiberites grambergi	Icriospathodus collinsoni	OI 4
		sn6		4 zones	2 zones	Neospath. pingdingshanensis	
	Olenekian			Anasiirites kingianus	Wassachites tardus		
				Meekoceras gracilitatis	Lepiskites kolymensis	3 zones	
	249.5			Flemingites flemingianus	Hedenstr. hedenstroemi		
250		sn3		R. rohilla - G. frequens	4 zones	Sweetospathodus kummeli	
				P. planidorsatus - Discophic.	Wordieoceras decipiens	Neogondolella krystyni	
	Induan			Ophiceras tibeticum	Tompophiceras morpheus	Isarcicella isarcica	
251.0		sn1		Otoc.woodwardi-fissisellatum	Tompophiceras pascoei	Hindeodus parvus	
				Hypophic. changxingense		Clarkina meishanensis	
	Permian			Pleuronodoc. occidentale	Otoceras boreale	Clarkina hauschkei	end-Perm

Figure 10.4. (cont.)

(e.g., Bowring, 2007, Carboniferous–Permian Congress, unpublished presentation). Second, combined magnetostratigraphy and cyclostratigraphy of the Buntsandstein can be used to scale the Early Triassic magnetostratigraphy from other regions (e.g., Szurlies, 2004, 2007) – this cycle scaling is consistent with the GTS04 estimates of the

Alternate Time Scale for Triassic

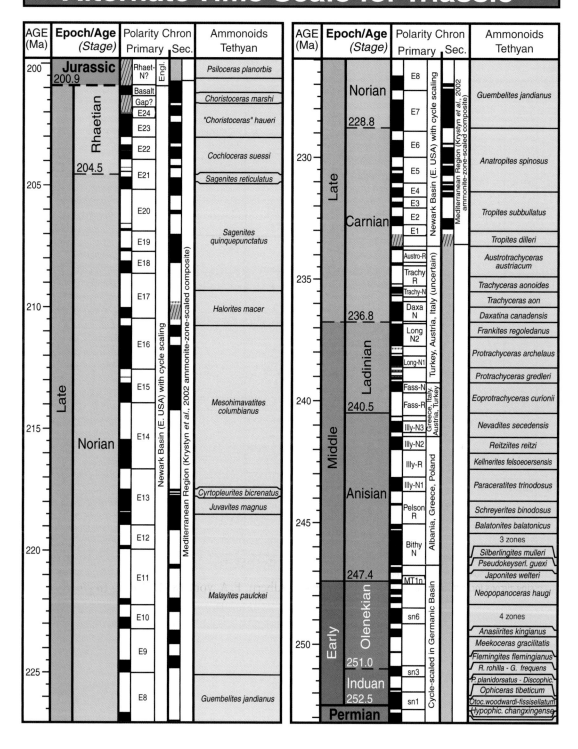

AGE (Ma)	Epoch/Age (Stage)	Polarity Chron Primary	Sec.	Ammonoids Tethyan
200	Jurassic 200.9	Rhaet-N?	Engl.	Psiloceras planorbis
	Rhaetian	Basalt Gap?		Choristoceras marshi
		E24		"Choristoceras" haueri
		E23		
		E22		Cochloceras suessi
205	204.5	E21		Sagenites reticulatus
		E20		Sagenites quinquepunctatus
		E19		
		E18		
	Late	E17		
210				Halorites macer
		E16		
	Norian	E15		Mesohimavatites columbianus
215		E14		
		E13		Cyrtopleurites bicrenatus
				Juvavites magnus
220		E12		
		E11		Malayites paulckei
		E10		
		E9		
225		E8		Guembelites jandianus

Newark Basin (E. USA) with cycle scaling
Mediterranean Region (Krystyn et al., 2002 ammonite-zone-scaled composite)

AGE (Ma)	Epoch/Age (Stage)	Polarity Chron Primary	Sec.	Ammonoids Tethyan
	Norian	E8		Guembelites jandianus
		E7		
	228.8	E6		Anatropites spinosus
230		E5		
		E4		
	Late	E3		Tropites subbullatus
	Carnian	E2		
		E1		Tropites dilleri
		Austro-R		Austrotrachyceras austriacum
		Trachy R		
235		Trachy-N		Trachyceras aonoides
		Daxa N		Trachyceras aon
	236.8			Daxatina canadensis
		Long N2		Frankites regoledanus
	Ladinian			Protrachyceras archelaus
		Long-N1		Protrachyceras gredleri
		Fass-N		Eoprotrachyceras curionii
240	240.5	Fass-R		
		Illy-N3		Nevadites secedensis
		Illy-N2		Reitziites reitzi
	Middle	Illy-R		Kellnerites felsoeoersensis
	Anisian	Illy-N1		Paraceratites trinodosus
		Pelson R		Schreyerites binodosus
245				Balatonites balatonicus
		Bithy N		3 zones
				Silberlingites mulleri
				Pseudokeyserl. guexi
	247.4	MT1n		Japonites welteri
				Neopopanoceras haugi
	Early	sn6		4 zones
250	Olenekian			Anasiirites kingianus
				Meekoceras gracilitatis
	251.0			Flemingites flemingianus
		sn3		R. rohilla - G. frequens
	Induan			P. planidorsatus - Discophic.
	252.5	sn1		Ophiceras tibeticum
	Permian			Otoc.woodwardi-fissisellatum
				Hypophic. changxingense

Newark Basin (E. USA) with cycle scaling
Mediterranean Region (Krystyn et al., 2002 ammonite-zone-scaled composite)
Turkey, Austria, Italy (uncertain)
Greece, Italy, Austria, Turkey
Albania, Greece, Poland
Cycle-scaled in Germanic Basin

durations of the Lower Triassic stages and has been incorporated into the Triassic time scales compiled here. Third, additional radiometric ages constrain the beginnings of the Anisian, Ladinian and Norian. Fourth, it now appears that the "Option #2 Long Norian" correlation of Muttoni *et al.* (2004) is a more appropriate fit to the Newark cycle-scaled magnetic pattern.

These advances imply that the duration of the Norian spans nearly half of the Triassic, the Anisian–Carnian interval is much shorter, and numerical ages for base Triassic through Anisian are shifted older by about 1.5 myr compared to GTS04. Because these new advances are so dramatic compared to any other period in the Phanerozoic, we have included a possible "Potential Triassic numerical scaling – version 2008" as a figure in this chapter (see Figure 10.5). This is broadly similar to the estimated ages for the Triassic stages by Brack *et al.* (2005). The numerical age and span of the Rhaetian Stage is still debated. The suggested

placement of Rhaetian magnetostratigraphy relative to Newark cycle-scaled magnetics shown in both figures in this chapter is modified from Gallet *et al.* (2007).

In addition to the current search for additional radiometric age constraints on the Triassic time scale, it is probable that additional cycle stratigraphy from central Europe and other regions will provide a more robust estimate of durations of biozones. Isotope stratigraphy and other correlation methods will aid in the inter-regional correlations, including marine–terrestrial ties.

Acknowledgements

Michael Orchard (Geological Survey of Canada, chair of Triassic Subcommission) contributed to this review. For further details/information, we recommend "The Triassic Period" by J. G. Ogg (in *A Geologic Time Scale 2004*), which had contributions from many specialists. Portions of the background material are from documents of the Triassic Subcommission.

Figure 10.5. Revised Triassic time scale (version 2007). Following the publication of GTS04, it became apparent from new methods of processing zircons for U-Pb age-dating, additional radiometric ages, and revised magnetostratigraphic correlations that the estimated age assignments of most Triassic stage and substage boundaries require major adjustments. This chart indicates how these ages might be adjusted (modified from Brack *et al.*, 2005). For comparison, the ammonoid zonation is the same as in Figure 9.4 (modified from Kozur, 2003 and pers. comm., 2006). It is probable that these age assignments will undergo further adjustment in the coming years, especially through the efforts of EarthTime to achieve inter-laboratory standardization and cycle-correlated biomagnetostratigraphy.

Further reading

Alberti, F. A. von, 1834. *Beitrag zu einer Monographie des Bunter Sandsteins, Muschelkalks und Keupers und die Verbindung dieser Gebilde zu einer Formation.* Stuttgart: J. G. Cottaíshen.

Bachmann, G. H., and Kozur, H. W., 2004. The Germanic Triassic: correlations with the international chronostratigraphic scale, numerical ages and Milankovitch cyclicity.

Hallesches Jahrbuchfür Geowissenschaften, **B26**: 17–62.

Bowring, S. A., Erwin, D. H., Jin, Y. G., Martin, M. W., Davidek, K., and Wang, W., 1998. U/Pb zircon geochronology and tempo of the end-Permian mass extinction. *Science*, **280**: 1039–1045.

Brack, P., Rieber, H., Nicora, A., and Mundil, R., 2005. The Global boundary Stratotype Section and Point (GSSP) of the Ladinian Stage (Middle Triassic) at Bagolino (Southern Alps, Northern Italy) and its implications for the Triassic time scale. *Episodes*, **28**(4): 233–244.

de Graciansky, P. -C., Hardenbol, J., Jacquin, Th., and Vail, P. R. (eds.), 1998. *Mesozoic–Cenozoic Sequence Stratigraphy of European Basins. SEPM Special Publication* 60.

Erwin, D. H., 2006. *Extinction: How Life on Earth Nearly Ended 250 Million Years Ago*. Princeton: Princeton University Press.

Furin, S., Preto, N., Rigo, M., Roghi, G., Gianolla, P., Crowley, J. L., and Bowring, S. A., 2007. High-precision U–Pb zircon age from the Triassic of Italy: implications for the Triassic time scale and the Carnian origin of calcareous nannoplankton and dinosaurs. *Geology*, **34**: 1009–1012.

Galfetti, T., Bucher, H., Brayard, A., Hochuli, P. A., Weissert, H., Guodun, K., Atudorei, V., and Guex, J., 2007. Late Early Triassic climate change: insights from carbonate carbon isotopes, sedimentary evolution and ammonoid paleobiogeography. *Palaeogeography, Palaeoclimatology, Palaeoecology*, **243**: 394–411.

Gallet, Y., Krystyn, L., Marcoux, J., and Bess, J., 2007. New constraints on the end-Triassic (Upper Norian–Rhaetian) magnetostratigraphy. *Earth and Planetary Science Letters*, **255**: 458–470.

Hardenbol, J., Thierry, J., Farley, M. B., Jacquin, Th., de Graciansky, P. -C., and Vail, P. R., 1998. Mesozoic and Cenozoic sequence chronostratigraphic framework of European basins. In: *Mesozoic–Cenozoic Sequence Stratigraphy of European Basins*, eds. P. -C. de Graciansky, J. Hardenbol, Th. Jacquin, and P. R. Vail. *SEPM Special Publication* **60**: 3–13, 763–781, and chart supplements.

Hounslow, M. W., Hu, M., Mørk, A., Weitschat, W., Vigran, J. O., Karloukovski, V., and Orchard, M. J., 2007a. Intercalibration of Boreal and Tethyan timescales: the magneto-biostratigraphy of the Botneheia Formation (Middle Triassic) and the late Early Triassic, Svalbard (arctic Norway): polar research. In: *The Global Triassic*, eds. S. G. Lucas and J. A. Spielmann. *New Mexico Museum of Natural History and Science Bulletin*, **41**: 68–70.

Hounslow, M. W., Szurlies, M., Muttoni, G., and Nawrocki, J., 2007b. The magnetostratigraphy of the Olenekian–Anisian boundary and a proposal to define the base of the Anisian using a magnetozone datum. *Albertiana*, **36**: 72–77.

Kent, D. V., and Olsen, P. E., 1999. Astronomically tuned geomagnetic polarity timescale for the Late Triassic. *Journal of Geophysical Research*, **104**: 12 831–12 841. [On-line update (2002) posted at Newark Basin Coring Project website, www.ldeo.columbia. edu/~polsen/nbcp/nbcp.timescale.htm]

Kozur, H. W., 2003. Integrated ammonoid-, conodont and radiolarian zonation of the Triassic. *Hallesches Jahrbuch für Geowissenschaften*, **B25**: 49–79.

Kozur, H. W., and Bachmann, G. H., 2005. Correlation of the Germanic Triassic with the international scale. *Albertiana*, **32**: 21–35.

Lucas, S. G., and Spielmann, J. A. (eds.), 2007. *The Global Triassic. New Mexico Museum of Natural History and Science Bulletin*, 41.

Menning, M., Gast, R., Hagdorn, H., Kading, K.-C., Simon, T., Szurlies, M., and Nitsch, E., 2005. Zeitskala für Perm und Trias in der Stratigraphischen Tabelle von Deutschland 2002, zyklostratigraphische Kalibrierung von höherer Dyas und Germanischer Trias und das Alter der Stufen Roadium bis Rhaetium 2005. In: *Erläuterungen zur Stratigraphischen Tabelle von Deutschland*, eds. M. Menning and A. Hendrich. *Newsletters of Stratigraphy*, **41**(1/3): 173–210.

Mundil, R., Ludwig, K. R., Metcalfe, I., and Renne, P. R., 2004. Age and timing of the Permian mass extinctions: U/Pb dating of closed-system zircons. *Science*, **305**: 1760–1763. [Note: This is one of several articles by different groups on Early Triassic isotopes.]

Muttoni, G., Kent, D. V., Olsen, P. E., Lowrie, W., Bernasconi, S. M., and Hernández, F. M., 2004. Tethyan magnetostratigraphy from Pizzo Mondello (Sicily) and correlation to the Late Triassic Newark astrochronological polarity time scale. *Geological Society of America Bulletin*, **116**: 1043–1058.

Orchard, M. J., 2007. A proposed Carnian-Norian Boundary GSSP at Black Bear Ridge, northeast British Columbia, and a new conodont framework for the boundary interval. *Albertiana* **36**: 130–141 (and references to other Orchard 2007 papers, therein)

Pálfy, J., Mortensen, J. K., Carter, E. S., Smith, P. L., Friedman, R. M., and Tipper, H. W., 2000. Timing the end-Triassic mass extinction: first on land, then in the sea? *Geology*, **28**: 39–42.

Pálfy, J., Parris, R. R., David, K., and Vörös, A., 2003. Mid-Triassic integrated U-Pb geochronology and ammonoid biochronology from the Balaton Highland (Hungary). *Journal of the Geological Society of London*, **160**: 271–284.

Szurlies, M., 2004. Magnetostratigraphy: the key to global correlation of the classic Germanic Trias – case study Volpriehausen Formation (Middle Buntsandstein), Central Germany. *Earth and Planetary Science Letters*, **227**: 395–410.

Szurlies, M., 2007. Latest Permian to Middle Triassic cyclo-magnetostratigraphy from the Central European Basin, Germany: implications for the geomagnetic polarity timescale. *Earth and Planetary Science Letters*, **261**: 602–619.

Tozer, E. T., 1984. *The Trias and its Ammonites: The Evolution of a Time Scale. Geological Survey of Canada Miscellaneous Report*, 35.

Selected on-line references

Triassic Subcommission – *paleo.cortland. edu/sts/*. The Albertiana newsletter of the subcommission has its own site – *www.bio.uu.nl/%7Epalaeo/Albertiana/ Albertiana01.htm*.

We recommend the extensive Triassic webpages and links at *Palaeos*, Smithsonian Institution, University of California Museum of Paleontology, and *Wikipedia*. See URL details at end of Chapter 1.

11
Jurassic Period

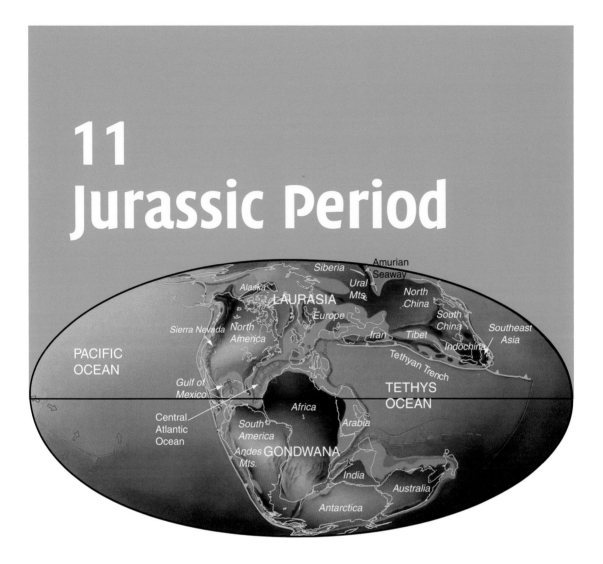

History and base of Jurassic

Figure 11.1. Geographic distribution of the continents during the Jurassic Period (152 Ma). The paleogeographic map was provided by Christopher Scotese.

The framework of the modern Jurassic that was established by Leopold von Buch (1839) built upon the concept of "Terrains Jurassique" of Alexander Brongniart (1829), which was named after the Jura Mountains of eastern France and western Switzerland. Alcide d'Orbigny grouped the Jurassic ammonite and other fossil assemblages of France and England into "étages" during 1842 through 1852, and seven of his ten Jurassic stages are used today, although none of them in their original stratigraphic range. An international consensus on the Jurassic stage nomenclature and general definitions was established by the International Subcommission on Jurassic Stratigraphy in 1962 and 1967. Formalizing the GSSPs for these international units proved more elusive.

The end-Triassic mass extinctions terminated many groups of marine life,

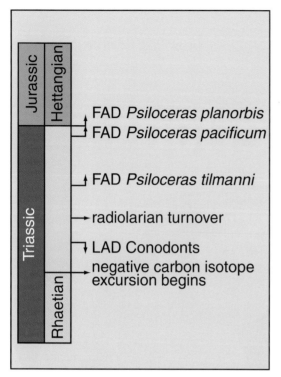

Figure 11.2. Succession of potential marker events for definition of the Triassic–Jurassic boundary (after Lucas *et al.*, in Lucas and Tanner, 2007).

including the conodonts which produced the distinctive phosphatic jaw elements that are used for high-resolution zonation of the upper Cambrian through Triassic. Most groups of ammonoids also vanished, and non-biotic forms of correlation are required for high-resolution correlation within the Triassic–Jurassic boundary interval. A negative carbon-isotope excursion coincides with the mass extinction, which may have been partially triggered by the extensive eruption of the Central Atlantic magmatic province at ~200 Ma.

The traditional base of the Jurassic was the appearance of the first ammonites (*Psiloceras* genus) above an ammonoid-barren interval, but it was discovered that different species of *Psiloceras* had diachronous appearances according to the paleogeographic region. The leading candidate for the base-Jurassic GSSP is at the Kuhjoch section, Tyrol, Austria, where the first occurrence of *Psiloceras spelae* is approximately coincident with the end of the latest-Triassic diversity crisis.

International subdivisions of Jurassic

The concept of biostratigraphic zones was first developed in ammonite-rich Jurassic strata, and ammonite zones remain the main method of relative dating and correlation within the Jurassic. Even though there is a standardized ammonite zonation of each European stage, the establishment of GSSPs required agreement by the Jurassic Subcommission working groups on the basal ammonite horizon within each zone and identification of secondary criteria for precise global correlation. This was particularly important for the Upper Jurassic, in which high-latitude Boreal ammonite zones had not been directly correlated to low-latitude Tethyan zones, and different stage definitions had been used. Therefore, the decision to establish the base-Kimmeridgian GSSP in the Boreal realm implied that approximately 1 ½ ammonite zones (~1 myr) of the traditional Oxfordian in the Tethyan realm have become part of the global Kimmeridgian.

Table 11.1 GSSPs of Jurassic stages, with location and primary correlation criteria (status in 2008)

Stage	GSSP location	Latitude, longitude	Boundary level	Correlation events	Reference
Tithonian	Candidates are Mt. Crussol or Canjuers, SE France, and Fornazzo, Sicily, S Italy			Near base of *Hybonoticeras hybonotum* ammonite zone and lowest occurrence of *Gravesia* genus, and the base of magnetic polarity chronozone M22An	
Kimmeridgian	Candidate is Flodigarry, Isle of Skye, NW Scotland			Ammonite, near base of *Pictonia baylei* ammonite zone of Boreal realm	
Oxfordian	Candidates are Savouron, Provence, SE France, and Redcliffe Point, Dorset, SW England			Ammonite *Cardioceras redcliffense* Horizon at base of the *Cardioceras scarburgense* Subzone (*Quenstedtoceras mariae* Zone)	
Callovian	Candidate is Pfeffingen, Swabian Alb, SW Germany			Ammonite, FAD of the genus *Kepplerites* (*Kosmoceratidae*) (defines base of *Macrocephalites herveyi* Zone in sub-Boreal province of Great Britain to SW Germany)	
Bathonian	Ravin du Bès, Bas-Auran area, Alpes de Haute Provence, France	43° 57' 38" N 6° 18' 55" E[a]	Base of limestone bed RB07	Ammonite, FAD of *Gonolkite convergens* (defines base of *Zigzagiceras zigzag* Zone)	
Bajocian	Murtinheira Section, Cabo Mondego, Portugal	40° 11' 57" N 8° 54' 15" W[a]	Base of Bed AB11 of the Murtinheira Section	Ammonite FAD *Hyperlioceras mundum*, *H. furcatum*, *Braunsina aspera*, *B. elegantula*	*Episodes* **20**(1), 1997
Aalenian	Fuentelsaz, Spain	41° 10' 15" N 1° 50' W	Base of Bed FZ 107 in Fuentelsaz Section	Ammonite FAD *Leioceras opalinum* and *L. lineatum*	*Episodes* **24**(3), 2001
Toarcian	Peniche, Portugal			Ammonite, near FAD of a diversified *Eodactylites* ammonite fauna; correlates with the NW European *Paltus* horizon	
Pliensbachian	Robin Hood's Bay, Yorkshire, England	54° 24' 25" N 0° 29' 51" W	Base of Bed 73b at Wine Haven, Robin Hood's Bay	Ammonite association of *Bifericeras donovani* and *Apoderoceras* sp.	*Episodes* **29**(2), 2006
Sinemurian	East Quantoxhead, SW England	51° 11' 27.3" N 3° 14' 11.2" W[a]	0.90 m above the base of Bed 145	Ammonite FAD *Vermiceras quantoxense*, *V. palmeri*	*Episodes* **25**(1), 2002
Hettangian (base Jurassic)	Candidate is Kuhjoch section, Tyrol, Austria	47° 29' 02" N 11° 31' 50" E		FAD of first species (*Ps. spelae*) of *Psiloceras* ammonite group; significantly above a sharp negative carbon-isotope excursion	

a. According to Google Earth.
Source: Details on each GSSP are available at *www.stratigraphy.org* and in the *Episodes* publications.

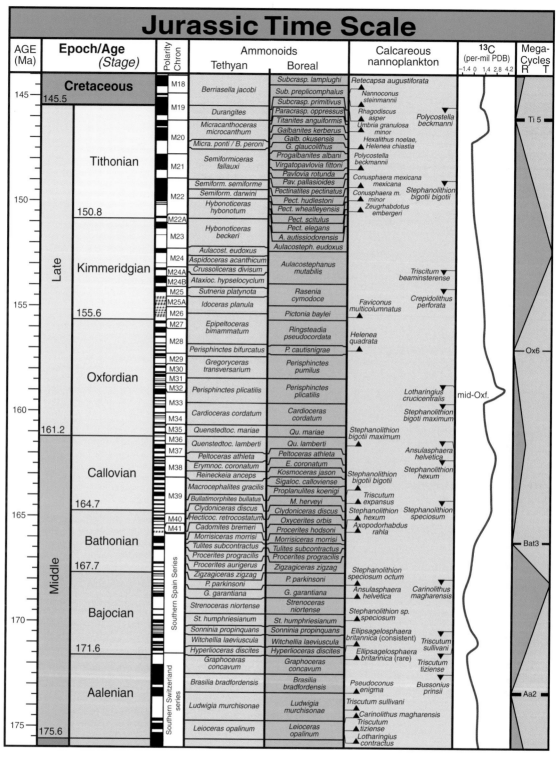

Figure 11.3. Numerical ages of epoch/series and age/stage boundaries of the Jurassic with major marine biostratigraphic zonations and principle eustatic trends. ["Age" is the term for the time equivalent of the rock-record "stage".] The ammonite scales are summarized from Groupe français d'étude du Jurassique (1997) and other sources. The calcareous nannoplankton scale was provided by Jim Bergen (BP), and was partially based on Bown and Cooper (1998). The ¹³C curve is generalized from Jenkyns *et al.* (2002) with additional details from Glowniak and Wierzbowski (2007) for middle Oxfordian, Kemp *et al.* (2005) for lower Toarcian, and Pálfy *et al.* (2001) for Triassic–Jurassic boundary interval. The Mega cycles are from Hardenbol *et al.* (1998).

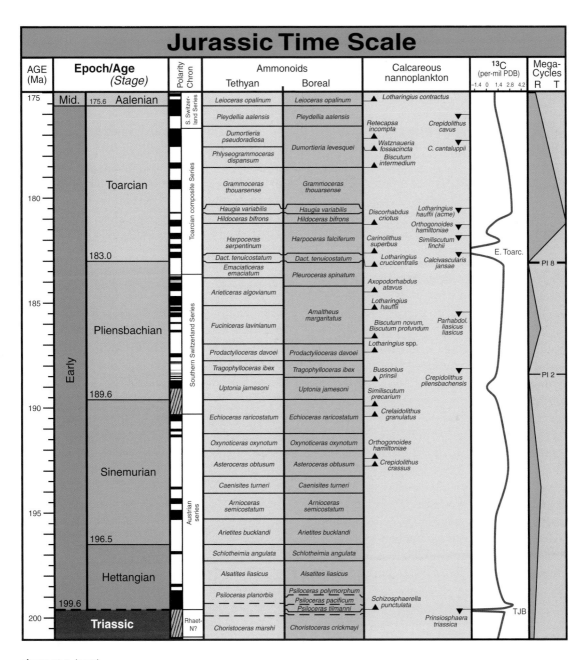

Figure 11.3. (cont.)

Selected aspects of Jurassic stratigraphy

Biostratigraphy

Biostratigraphic zones are named according to the genus–species of associated taxa (e.g., *Pictonia baylei* zone). Ammonite workers in the Jurassic often use a "standard zone" that is not directly associated with the biotic range of the species that lends its name and is indicated by a non-italicized name (e.g., Baylei Zone; in which the actual regional occurrence of *P. baylei* may be quite high within the zone). This use of ammonite-named "standard zones" appears to be unique to the Jurassic. In the time-scale figures for this chapter, we have written the ammonite zones with *genus–species* as is required for biostratigraphy, but the actual definition of those zones may have no correspondence to the ranges of the taxa. Yes, it is confusing.

The Jurassic seas saw an increase in abundance of planktonic tests, which become a major or dominant constituent of deep-water sediments. Calcareous nannoplankton, organic-walled dinoflagellates and siliceous radiolarians are well-developed biostratigraphic tools for oceanic sediments.

Dinosaurs are the most famous Jurassic fauna, but the biostratigraphic ranges of these enormous creatures are not yet well established relative to marine-based stages.

Stable-isotope stratigraphy

Major negative carbon-isotope excursions occur in the Triassic–Jurassic boundary interval, the lowest Toarcian and the middle Oxfordian. The lower Toarcian excursion, one of the largest in the Phanerozoic, is associated with an increase in the strontium and osmium isotopic ratios. The Toarcian event is coincident with the eruption of the Karroo–Ferrar flood basalts across South Africa and Antarctica.

Magnetic stratigraphy

The oldest preserved Pacific crust is Bajocian. Deep-tow magnetometer surveys of the former "Jurassic Quiet Zone" of Bajocian through Oxfordian has revealed numerous close-spaced "pre-M25" magnetic anomalies, which is consistent with compilations of magnetostratigraphy of sedimentary sections. However, the high frequency and lack of a distinctive "fingerprint" of these geomagnetic reversals has hindered their use in global correlation. From Kimmeridgian through Aptian, the M-sequence of the relatively longer-duration M25 through M0r chrons is a powerful correlation tool and a means for deriving relative durations of the biotic zones within magnetostratigraphic sections.

Cycle and magnetic stratigraphy

Several intervals within the Jurassic have been scaled according to interpretations of the orbital-climate Milankovitch cyclicity within the associated sediments. These studies, which were partially incorporated in scaling biozones in the Jurassic time scale, await independent confirmation.

Numerical time scale (GTS04 and future developments)

Direct radiometric dating within the Jurassic is available only for a few intervals (e.g., Triassic–Jurassic boundary interval, Toarcian, upper Oxfordian). Therefore, interpolation of ages for biozones and stage boundaries within *A Geologic Time Scale 2004* relied on a combination of assumptions of linearity of strontium-isotope trends, durations from cycle stratigraphy, constant Pacific spreading rates for magnetic anomalies, and relative numbers of ammonite subzones.

The Late Jurassic estimates from correlation to "constant spreading" magnetic anomalies have proven consistent with post-2004 publication of new radiometric ages (e.g., a 144.6 ± 0.8 Ma age for earliest Berriasian: Mahoney *et al*., 2005; and 154.1 ± 2.2 Ma for the base-Kimmeridgian GSSP: Selby, 2007). Cycle-derived durations for Kimmeridgian ammonite zones are also consistent with the GTS04 estimates. The heavily interpolated mid-Jurassic time scale awaits confirmation.

Acknowledgements

For further details/information, we recommend "The Jurassic Period" by J. G. Ogg (in *A Geologic Time Scale 2004*) which had contributions from many specialists. Portions of the background material are from documents of the Jurassic Subcommission.

Further reading

Special Jurassic issue of *Proceedings of the Geologists' Association* (2008; **119**; 1–117), including articles on GSSPs by Nicol Morton (96–104), time scale by József Pálfy (85–95), ammonoids by Kevin Page (35–57), sequence stratigraphy by Stephen Hesselbo (19–34), and Jurassic–Cretaceous boundary by John Cope (105–117).

Arkell, W. J., 1956. *Jurassic Geology of the World*. Edinburgh: Oliver & Boyd.

Bown, P. R., and Cooper, M. K. E., 1998. Jurassic. In: *Calcareous Nannofossil Biostratigraphy*, ed. P. R. Bown. London: Kluwer Academic Publishers, pp. 34–85.

Brongniart, Q., 1829. *Tableau des terrains qui composent l'écorce du globe ou Essai sur la structure de la partie connue de la terre*. Paris.

de Graciansky, P.-C., Hardenbol, J., Jacquin, Th., and Vail, P. R (eds.), 1998. *Mesozoic–Cenozoic Sequence Stratigraphy of European Basins*, SEPM Special Publication no. 60. Tulsa.

Glowniak, E., and Wierzbowski, H., 2007. Comment on "The mid-Oxfordian (Late Jurassic) positive carbon-isotope excursion recognised from fossil wood in the British Isles" by C. R. Pearce, S. P. Hesselbo, A. L. Coe, Palaeogeography, Palaeoclimatology, Palaeoecology, **221**: 343–357. *Palaeogeography, Palaeoclimatology, Palaeoecology*, **248**: 252–254.

Groupe français d'étude du Jurassique (coordinated by Cariou, E., and Hantzpergue, P.), 1997. Biostratigraphie du Jurassique ouest-européen et méditerranéen: zonations parallèles et distribution des invertébrés et microfossiles. *Bulletin des Centres de Recherches Exploration – Production Elf-Aquitaine, Mémoire*, **17**.

Hardenbol, J., Thierry, J., Farley, M. B., Jacquin, Th., de Graciansky, P.-C., and Vail, P. R. (with numerous contributors), 1998. Mesozoic and Cenozoic sequence chronostratigraphic framework of European basins. In: *Mesozoic–Cenozoic Sequence Stratigraphy of European Basins*, eds. P.-C. de Graciansky, J. Hardenbol, Th. Jacquin, and P. R. Vail. *SEPM Special Publication*, **60**: 3–13, 763–781, and chart supplements.

Hesselbo, S. P., McRoberts, C. A., and Palfy, J., 2007. Triassic–Jurassic boundary events: problems, progress, possibilities. *Palaeogeography, Palaeoclimatology, Palaeoecology*, **244**: 1–10.

Jenkyns, H. C., Jones, C. E., Gröcke, D. R., Hesselbo, S. P., and Parkinson, D. N., 2002. Chemostratigraphy of the Jurassic System: applications, limitations and implications for palaeoceanography. *Journal of the Geological Society of London*, **159**: 351–378.

Kemp, D. B., Coe, A. L., Cohen, A. S. and Schwark, L., 2005. Astronomical pacing of methane release in the Early Jurassic. *Nature*, **437**: 396–399

Lucas, S. G., and Tanner, L. H., 2007. The nonmarine Triassic–Jurassic boundary in the Newark Supergroup of eastern North America. *Earth-Science Review*, **84**: 1–20.

Mahoney, J. J., Duncan, R. A., Tejada, M. L. G., Sager, W. W., and Bralower, T. J., 2005. Jurassic–Cretaceous boundary age and mid-ocean-ridge-type mantle source for Shatsky Rise. *Geology*, **33**: 185–188.

Morton, N., 2006. Chronostratigraphic units in the Jurassic and their boundaries: definition, recognition and correlation, causal mechanisms. In: *Marine and Non-Marine Jurassic: Boundary Events and Correlation*, eds. J. Sha, Y. Wang, and S. Turner, *Progress in Natural Science*, **16** (Special Issue): 1–11.

Ogg, J. G., Karl, S. M., and Behl, R. J., 1992. Jurassic through Early Cretaceous sedimentation history of the central Equatorial Pacific and of Sites 800 and 801. *Proceedings Ocean Drilling Program, Scientific Results*, **129**: 571–613.

Pálfy, J., 2008. The quest for refined calibration of the Jurassic time scale. *Proceedings of the Geologists' Association*, **119**: 85–95.

Pálfy, J., Demény, A., Haas, J., Hetényi, M., Orchard, M., and Vetö, I., 2001. Carbon isotope anomaly and other geochemical changes at the Triassic–Jurassic boundary from a marine section in Hungary. *Geology*, **29**: 1047–1050.

Selby, D., 2007. Direct rhenium–osmium age of the Oxfordian–Kimmeridgian boundary, Staffin Bay, Isle of Skye, UK, and the Late Jurassic time scale. *Norwegian Journal of Geology*, **47**: 291–299.

von Buch, L., 1839, *Über den Jura in Deutschland*. Berlin: Königlich preussischen Akademie der Wissenschaften.

Selected on-line references

Jurassic Subcommission – *www.es.ucl.ac.uk/ people/bown/ISJSwebsite.htm* – contains GSSP information, newsletter copies, and links to other sites.

Jurassic Coast of Dorset and East Devon (UNESCO World Heritage site) – *www. jurassiccoast.com/index.jsp* – England's renowned deposits.

Jurassic Reef Park (by Reinhold Leinfelder, now director of Humboldt Museum of Natural History, Berlin) – *www.palaeo.de/ edu/JRP/Jurassic_Reef_Park.html* – a trip to Jurassic reefs with implications for present ones.

We recommend the extensive Jurassic webpages and links at *Palaeos*, Smithsonian Institution, University of California Museum of Paleontology, and *Wikipedia*. See URL details at end of Chapter 1.

12
Cretaceous Period

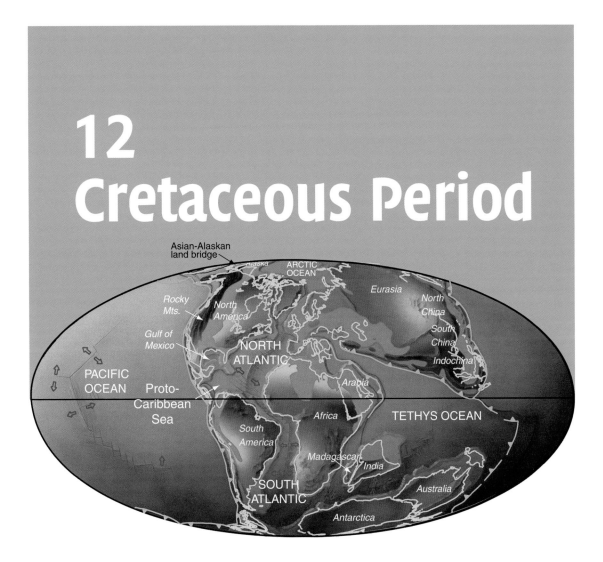

Asian-Alaskan land bridge

ARCTIC OCEAN

Alaska

Eurasia

Rocky Mts.

North America

North China

South China

Gulf of Mexico

NORTH ATLANTIC

Indochina

PACIFIC OCEAN

Proto-Caribbean Sea

Arabia

Africa

TETHYS OCEAN

South America

Madagascar

India

SOUTH ATLANTIC

Australia

Antarctica

History and base of Cretaceous

D'Omalius d'Halloy (1822) defined the Terrain Crétacé to include "the formation of the chalk (*creta* in Latin), with its tufas, its sands and its clays." Alcide d'Orbigny (1840) grouped the Cretaceous fossil assemblages of France into five "étages," and these stages underwent subsequent revision and enhancement over the next century to create the current 12 Cretaceous stages.

    Defining the base of the Cretaceous System was obscured by the addition of the lowermost Berriasian Stage, which partially overlapped the original concept of the Tithonian

Figure 12.1. Geographic distribution of the continents during the Cretaceous Period (94 Ma). The paleogeographic map was provided by Christopher Scotese.

Stage (uppermost Jurassic) and incorporated the lower portion of the original Valanginian. The "traditional" placement of the base of this Berriasian (hence, base Cretaceous) is in a relatively homogeneous interval in which no major evolutionary datum of any marine group, no geochemical excursion or no distinctive magnetic reversal occurs to provide a useful global marker for precise placement. The extreme provincialism of most marine organisms during this interval complicates the situation. No GSSP level has yet been designated (as of early 2008), but the general usage of the Jurassic–Cretaceous boundary is the cluster of the lowest *Berriasella jacobi* ammonite (sub-Mediterranean realm), base of Calpionellid Zone B (in low-latitude open marine deposits), and the later part of polarity Chron M19n.

International subdivisions of Cretaceous

Ammonites are the traditional means to subdivide the Cretaceous in each paleogeographic realm, but their provincialism during most of the Cretaceous has hindered the establishment of a global chronostratigraphy. Therefore, the ratified or candidate GSSPs established by the Cretaceous Subcommission have marine microfossils, benthic bivalves, magnetic reversals and carbon-isotope events as primary and secondary criteria for inter-regional correlation of the global stages.

Selected aspects of Cretaceous stratigraphy

Biostratigraphy

Ammonites and their belemnite cousins provided a powerful biostratigraphy for subdividing and correlating Cretaceous shelf successions within each region, but are restricted geographically. Therefore, most inter-regional biostratigraphic

correlations of marine strata rely on tests of carbonate (planktonic foraminifers, calpionellids, calcareous nannofossils) and organic (cysts of dinoflagellates) organisms. Calpionellids are enigmatic pelagic microfossils that appeared in the latest Jurassic and vanished in the latest Valanginian. The demise of calpionellids coincided with the increased abundance of planktonic foraminifers, which provide a powerful biostratigraphic tool in all ocean basins through to the Present. Calcareous nannoplankton, while useful in Jurassic stratigraphy, explode in abundance at the onset of the Cretaceous as the main contributor to the characteristic chalk deposits. Dinoflagellate cysts are particularly important in correlating organic-rich marine deposits. Silica-test radiolarians are of lesser importance than in the Jurassic.

Primary correlation markers for three of the proposed GSSPs of the Upper Cretaceous are from planktonic-dwelling crinoids and bottom-dwelling inoceramid bivalves. These macrofossils provide major regional zonations, especially within chalk deposits.

The land fauna was dominated by the dinosaurs and their bird relatives, but their stratigraphic record is poorly constrained. Mammals were always present, but only enable a broad zonation beginning in the uppermost Cretaceous.

Magnetic stratigraphy

The M-sequence of Late Jurassic and Early Cretaceous marine magnetic anomalies has been

Table 12.1 GSSPs of Cretaceous stages, with location and primary correlation criteria (status in 2008)

Stage	GSSP location	Latitude, longitude	Boundary level	Correlation events	Reference
Maastrichtian	Tercis les Bains, Landes, France	43° 40′ 46.1″ N 1° 06′ 47.9″ W[a]	Level 115.2 on platform IV of the geological site at Tercis les Bains	Mean of 12 biostratigraphic criteria of equal importance. Closely above is ammonite FAD of *Pachydiscus neubergicus*. Boreal proxy is belemnite FAD of *Belemnella lanceolata*	*Episodes* **24**(4), 2001
Campanian	Candidates are in southern England and in Texas			Crinoid extinction of *Marsupites testudinarius*	
Santonian	Leading candidates are Olazagutia, Spain, and Ten-Mile Creek, Texas			Inoceramid bivalve FAD of *Cladoceramus undulatoplicatus*	
Coniacian	Leading candidates are in Poland (Slupia Nadbrzena), USA (Pueblo, Colorado), and Germany (Salzgitter-Salder Quarry)			Inoceramid bivalve FAD of *Cremnoceramus rotundatus* (sensu Tröger non Fiege)	
Turonian	Pueblo, Colorado, USA	38° 16′ 56″ N 104° 43′ 39″ W[a]	Base of Bed 86 of the Bridge Creek Limestone Member	Ammonite FAD *Watinoceras devonense*	*Episodes* **28**(2), 2005
Cenomanian	Mount Risou, Hautes-Alpes, France	44° 23′ 33″ N 5° 30′ 43″ E	36 m below the top of the Marnes Bleues Formation on the south side of Mont Risou	Planktonic foraminifer FAD *Rotalipora globotruncanoides*	*Episodes* **27**(1), 2004
Albian	Southeastern France			Candidates include: (1) calcareous nannofossil FAD of *Praediscosphaera columnata* (= *P. cretacea* of some earlier studies), (2) carbon-isotope excursion (black-shale episode), (3) ammonite	
Aptian	Candidate is Gorgo a Cerbara, Piobbico, Umbria-Marche, central Italy			Magnetic polarity chronozone, base of M0r; near base of *Paradeshayesites oglanlensis* ammonite zone	
Barremian	Candidate is Río Argos near Caravaca, Murcia Province, Spain			Ammonite FAD of *Spitidiscus hugii – Spitidiscus vandeckii* group	
Hauterivian	Candidate is La Charce village, Drôme Province, SE France			Ammonite FAD of genus *Acanthodiscus* (especially A. *radiatus*)	
Valanginian	Candidates are near Montbrun-les-Bains, Drôme Province, SE France, and Cañada Luenga, Betic Cordillera, S Spain			Calpionellid FAD of *Calpionellites darderi* (base of Calpionellid Zone E); followed by the FAD of ammonite "*Thurmanniceras*" *pertransiens*	
Berriasian (base Cretaceous)				Maybe near ammonite FAD of *Berriasella jacobi*	

a. According to Google Earth.
Source: Details on each GSSP are available at *www.stratigraphy.org* and in the *Episodes* publications.

correlated to several ammonite, calpionellid, calcareous nannofossil and other biostratigraphic datums. The M-sequence pattern in the Pacific is the reference scale for relative durations of the magnetic chrons and the calibrated biostratigraphic events. The youngest well-documented M-sequence reversed-polarity anomaly is the brief M0r, which is the proposed main global marker for the base of the Aptian Stage.

The post-M0r "Cretaceous Quiet Zone" of constant normal polarity precludes usage of magnetostratigraphy from Aptian through Santonian. The base of the Campanian coincides with the onset of magnetic reversals of the C-sequence of marine anomalies, which continue to the Present.

Stable-isotope stratigraphy and anoxic events

There are seven "named" positive excursions in carbon-13 in the Cretaceous. Several of these events are associated with anomalous abundances of organic-rich sediments or other evidence of low-oxygen conditions in mid-depth ocean to shelf settings. Some of these "oceanic anoxic events" (OAEs) appear to be the consequence of episodes of massive volcanic activity. For example, the early Aptian "OAE1a" and carbon-13 spike coincides with the development of the enormous Ontong Java Plateau flood basalts at about 120 Ma. At higher resolution, there are numerous low-amplitude carbon-isotope events that are identified within English and Italian chalk deposits that enable a precise inter-regional correlation.

Strontium isotopes have a progressive rise in the Berriasian through Barremian stages and the Coniacian through Maastrichtian stages. The independent calibration of regional successions of ammonites and other fauna to the global strontium curve has enabled inter-calibration of portions of the Tethyan, Boreal and North American Interior Seaway zonal schemes.

Although the Cretaceous is generally considered to have been a relatively warm interval in Earth history with elevated levels of atmospheric carbon dioxide, there is evidence of high-latitude freezing or even glacial activity. From Aptian through Campanian, the elevated carbon dioxide levels and sequestering of calcium carbonate in the flooded continental interior basins resulted in an enhanced carbonate dissolution in deeper waters. This elevated carbonate compensation depth (CCD) precluded preservation of limestone deposits on much of the ocean floor until the CCD descended significantly after the early Campanian.

Cycle and sequence stratigraphy

The carbonate-rich facies of oceanic and interior seaway settings were particularly responsive to orbital-climate oscillations during the Cretaceous. Analysis of Milankovitch-cycle signatures in these deposits is one of the main tools for determining relative durations of biozones, and is a constraint on determining spreading rates of the M-sequence of marine magnetic anomalies. Such cycle durations will become the main high-resolution constraint on the Cretaceous interval, and will soon link to the Cenozoic astronomical-tuned scale.

Marginal marine to deep-shelf successions are punctuated by transgressive and regressive episodes. The main sequence boundaries in the Aptian–Albian have been correlated to emergent horizons in carbonate caps to Pacific seamounts, which imply that some of these reflect global eustatic sea-level oscillations. The sequences are superimposed on a broad flooding of the continental margins and interiors that peaked in the early Late Cretaceous.

Numerical time scale (GTS04, corrections, and future developments)

A Geologic Time Scale 2004 used three methods to interpolate numerical ages for Cretaceous stage boundaries and other events. The correlation of biostratigraphic datums to a simple spreading model for the M-sequence magnetic polarity pattern is the primary standard for the Oxfordian through earliest Aptian interval. The entire Upper Cretaceous (Cenomanian through Maastrichtian) has the distinction of having the greatest frequency of precise radiometric ages on ammonite-zoned strata in the entire Phanerozoic, which enabled a statistical scaling of the North American ammonite zonation. The intervening Aptian–Albian was calibrated via cycle-derived durations of Albian and lowest Aptian

foraminifer zones and an arbitrary assignment of equal duration to ammonite subzones for the remainder of the Aptian. All other ages relied on correlations to these primary reference scales.

Base of Hauterivian is corrected to 133.9 Ma

In GTS04, the base of the Hauterivian Stage has been assigned an age of 136.4 Ma based on the reported occurrence of its earliest ammonites to near the beginning of magnetic polarity Chron M11n (Channell *et al.*, 1995). However, the GTS04 authors and reviewers were not aware that this published correlation had been revised. The base of the Hauterivian is now considered to be near the base of Chron M10n (e.g., Weissert *et al.*, 1998), as supported by later studies (e.g., McArthur *et al.*, 2007), and consistent with the cycle-scaled duration of the Valanginian. Maintaining the same M-sequence age assignments implies that the base of the Hauterivian is approximately 133.9 Ma. This revision, plus the recommended relative durations of Valanginian–Hauterivian ammonite zones (McArthur *et al.*, 2007) is shown in the Cretaceous time-scale diagram in this book. All associated Valanginian–Hauterivian events have been adjusted accordingly. [Helmut Weissert and John McArthur contributed this correction.]

Base of Coniacian is corrected to 88.6 Ma

The base of the Coniacian Stage is the lowest occurrence of *Cremnoceramus rotundatus*

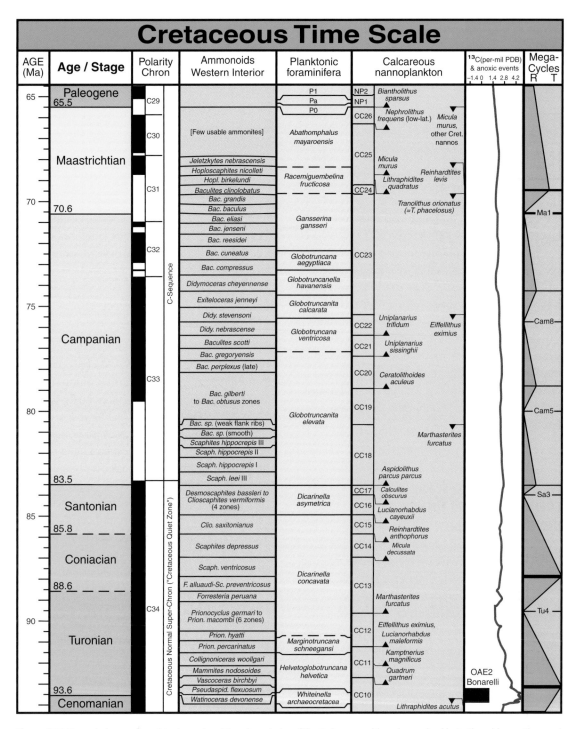

Figure 12.2. Numerical ages of epoch/series and age/stage boundaries of the Cretaceous with major marine biostratigraphic zonations and principle eustatic trends. ["Age" is the term for the time equivalent of the rock-record "stage".] Biostratigraphic scales include ammonoid, foraminifer, and calcareous nannoplankton zonations. The Western Interior and Tethyan ammonoid scales are respectively from Cobban and from Thierry *et al.* (both in Hardenbol *et al.*, 1998), with GTS04 revisions. The planktonic foraminifer and calpionellid scales are modified from ODP Leg 171B explanatory notes and from Robaszynski (in Hardenbol *et al.*, 1998) with partial recalibrations provided by Paul Sikora (EGI). The upper Cretaceous calcareous nannoplankton scale is modified from Erba *et al.* (1995) as tabulated in ODP Leg 171B Init. Repts. (Table 2, pp. 17–18). Lower and mid-Cretaceous nannoplankton zonations were compiled by Jim Bergen, based on publications by Tim Bralower *et al.* (1995), J. Bergen (1994) and information from Eric Kanael. The ^{13}C curve is generalized from Jarvis *et al.* (2006) for upper Cretaceous, from Föllmi *et al.* (2006) for lower Cretaceous, and from other sources. The Mega cycles are from Hardenbol *et al.* (1998).

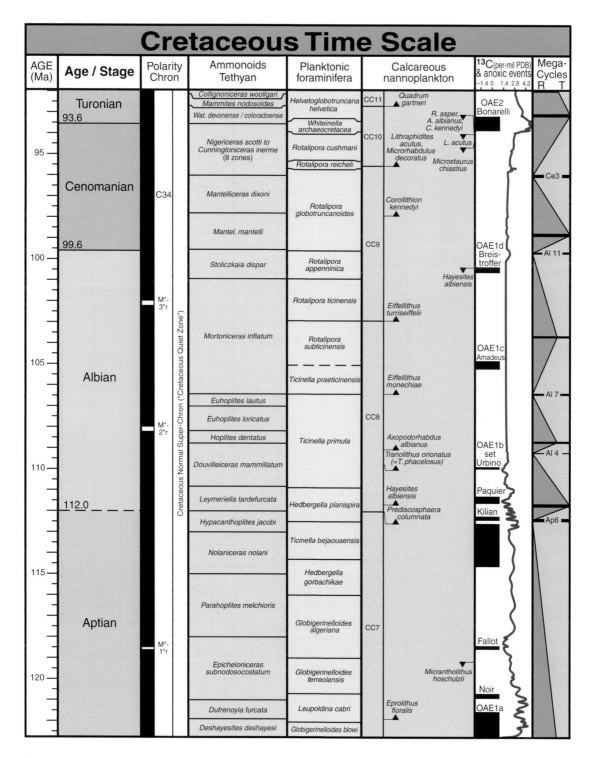

Figure 12.2. (cont.)

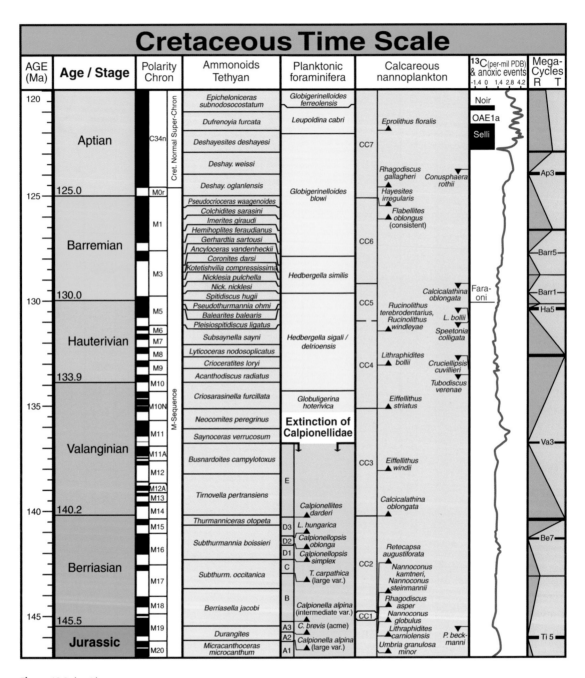

Figure 12.2. (cont.)

(sensu Tröger non Fiege), an inoceramid bivalve. In GTS04, this inoceramid datum had been considered to be slightly older than the lowest occurrence of ammonite *Forresteria petrocoriensis*, which had a spline-fit age of 89.07 Ma. Therefore, the base Coniacian had been estimated as 89.3 Ma. However, it is now known from inoceramid–ammonite calibration studies that the base of the *F. petrocoriensis* zone (or *Forresteria peruana* zone) is in upper Turonian, and that the Coniacian as defined above begins approximately at the base of the overlying *Scaphites preventricosus* ammonite zone (age of 88.58 Ma in GTS04, which had been used as the Lower/Middle Coniacian boundary level). Other than the inoceramid datum and associated base-Coniacian GSSP age, this correction does not alter any of the assigned ages for any other biostratigraphic or other stratigraphic events. [Ireneusz Walaszczyk, chair of Coniacian working group, contributed this correction.]

The Late Cretaceous time scale relies entirely on an extensive suite of Ar–Ar ages. The recalibration of the Ar–Ar monitor that is being considered by radiometric specialists, as discussed in the introductory chapter, will potentially shift all ages to an older date by about 1 myr. In the Early Cretaceous, a single U–Pb age on a magnetic reversal on a seamount constrains the entire scale, and acquiring additional radiometric ages is crucial. An age near the base of the Cretaceous (144.6 ± 0.8 Ma age for lowest Berriasian: Mahoney *et al.*, 2005) is consistent with the present M-sequence spreading model. It is probable that Milankovitch cycle-scaling of the Cretaceous will provide a precise scale in the near future.

Acknowledgements

For further details/information, we recommend "The Cretaceous Period" by J. G. Ogg, F. P. Agterberg, and F. M. Gradstein (in *A Geologic Time Scale 2004*), which had contributions from many specialists. Portions of the background material are from documents of the Cretaceous Subcommission.

Further reading

Bergen, J. A., 1994. Berriasian to early Aptian calcareous nannofossils from the Vocontian trough (SE France) and Deep Sea Drilling Site 534: new nannofossil taxa and a summary of low-latitude biostratigraphic events. *Journal of Nannoplankton Research (International Nannoplankton Association Newsletter)*, **16**: 59–69.

Bralower, T. J., Leckie, R. M., Slliter, W. V., and Thiestein, H. R., 1995. An integated Cretaceous microfossil biostratigraphy. In: *Geochronology, Time Scales and Global Stratigraphic Correlations: A Unified Temporal Framework for a Historical Geology*, eds. W. A. Berggren, D. V. Kent, and J. Hardenbol. *SEPM Special Volume*, **54**: 65–79.

Channell, J. E. T., Cecca, F., and Erba, E., 1995. Correlations of Hauterivian and Barremian (Early Cretaceous) stage boundaries to polarity chrons. *Earth and Planetary Science Letters*, **134**: 125–140.

de Graciansky, P.-C., Hardenbol, J., Jacquin, Th., and Vail, P. R. (eds.), 1998. *Mesozoic–Cenozoic Sequence Stratigraphy of European Basins. SEPM Special Publication*, **60**.

d'Halloy, J. G. J. d'O., 1822. Observations sur un essai de carte géologique de la France, des Pays-Bas, et des contrées voisines. *Annales des Mines*, **7**: 353–376.

d'Orbigny, A., 1840. *Paléontologie française. Terrains crétacés. I. Céphalopodes*. Paris.

Erba, E., Premoli Silva, I., and Watkins, D. K., 1995. Cretaceous calcareous plankton biostratigraphy of Sites 872 through 879. *Proceedings of the Ocean Drilling Program, Initial Reports*, **144**: 157–169.

Föllmi, K. B., Godet, A., Bodin, S., and Linder, P., 2006. Interactions between environmental change and shallow water carbonate buildup along the northern Tethyan margin and their impact on the Early Cretaceous carbon isotope record. *Paleoceanography*, **21**: PA4211, doi: 10.1029/2006PA001313.

Hardenbol, J., Thierry, J., Farley, M. B., Jacquin, Th., de Graciansky, P.-C., and Vail, P. R. (with numerous contributors), 1998. Mesozoic and Cenozoic sequence chronostratigraphic framework of European basins. In: *Mesozoic–Cenozoic Sequence Stratigraphy of European Basins*, eds. P.-C. de Graciansky, J. Hardenbol, Th. Jacquin, and P. R. Vail.

SEPM Special Publication **60**: 3–13, 763–781, and chart supplements.

Jarvis, I., Gale, A. S., Jenkyns, H. C., and Pearce, M. A., 2006. Secular variation in Late Cretaceous carbon isotopes: a new $\partial^{13}C$ carbonate reference curve for the Cenomanian–Campanian (99.6–70.6 Ma). *Geological Magazine*, **143**: 561–608.

Larson, R. L., and Erba, E., 1999. Onset of the mid-Cretaceous greenhouse in the Barremian–Aptian: igneous events and the biological, sedimentary, and geochemical responses. *Paleoceanography*, **14**: 663–678.

Mahoney, J. J., Duncan, R. A., Tejada, M. L. G., Sager, W. W., and Bralower, T. J., 2005. Jurassic–Cretaceous boundary age and mid-ocean-ridge-type mantle source for Shatsky Rise. *Geology*, **33**: 185–188.

McArthur, J. M., Janssen, N. M. M., Reboulet, S., Leng, M. J., Thirlwall, M. F., and van de Schootbrugge, B., 2007. Palaeotemperatures, polar ice-volume, and isotope stratigraphy (Mg/Ca, $\delta^{18}O$, $\delta^{13}C$, $^{87}Sr/^{86}Sr$): the Early Cretaceous (Berriasian, Valanginian, Hauterivian). *Palaeogeography, Palaeoclimatology, Palaeoecology*, **248**: 391–430.

ODP Leg 171B Scientific Party, 1998. Explanatory notes. *Proceedings of the Ocean Drilling Program, Initial Reports*, **171B**: 11–34. Available on-line at www-odp.tamu.edu/publications/171B_IR/CHAP_02.PDF

Weissert, H., Lini, A., Föllmi, K. B., and Kuhn, O., 1998. Correlation of Early Cretaceous carbon isotope stratigraphy and platform drowning events: a possible link? *Palaeogeography, Palaeoclimatololgy, Palaeoecology*, **137**: 189–203.

Wortmann, U. G., Jens Olaf Herrle, J. O., and Weissert, H., 2004. Altered carbon cycling and coupled changes in Early Cretaceous weathering patterns: evidence from integrated carbon isotope and sandstone records of the western Tethys. *Earth and Planetary Science Letters*, **220**: 69–82.

Selected on-line references

All Things Cretaceous (by Jen Aschoff, Montana State University) – *serc.carleton.edu/ research_education/cretaceous/index.html* – a digital resource collection for teaching and learning, as part of the DLESE Community Services Project.

Cretaceous Fossils (by Keith Minor) – *cretaceousfossils.com* – commercial site, but striving to document many fossils, especially ammonites, from the Cretaceous Period worldwide and make the information available to the amateur paleontology community.

British Chalk Fossils (by Robert Randell) – *www.chalk.discoveringfossils.co.uk* – chalk stratigraphy and images of main macrofossils.

We recommend the extensive Cretaceous webpages and links at *Palaeos*, Smithsonian Institution, University of California Museum of Paleontology, and *Wikipedia*. See URL details at end of Chapter 1.

13
Paleogene Period

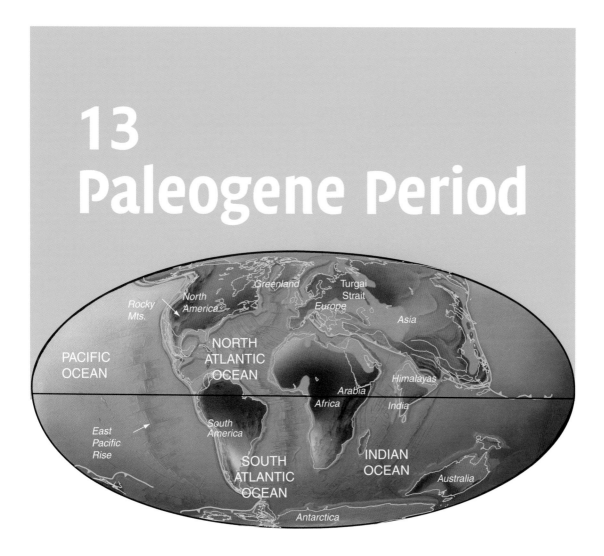

History and base of Paleogene

Figure 13.1. Geographic distribution of the continents during the Paleogene Period (50.2 Ma). The paleogeographic map was provided by Christopher Scotese.

The interval of the Cenozoic Era (originally "*Cainozoic*", from Greek *kainos* = new and *zoon* = animal) has undergone a complex history of alternate subdivisions. The seven Cenozoic epochs of Paleocene, Eocene, Oligocene, Miocene, Pliocene, Pleistocene, and Holocene epochs (respectively "old," "dawn," "few," "less," "more," "most" or "entire" + "new") were poetically named for relative abundances of modern forms among the fossil shells. A grouping of trios of these epochs into two periods of *Nummulitique* and *Neogénique*

was recommended in 1894 to the International Geological Congress. The same grouping of epochs (Paleocene through Oligocene, and Miocene through Holocene) was retained in the Paleogene (*palaios* = old, *genes* = born or clan) and in the Neogene (*neo* = new) plus Quaternary periods of the international scale.

The Paleogene Period/System was ratified in 1991 by IUGS upon the acceptance of the

Figure 13.2. The GSSP marking the base of the Paleogene System and its lowermost Danian Stage at El Kef, Tunisia.

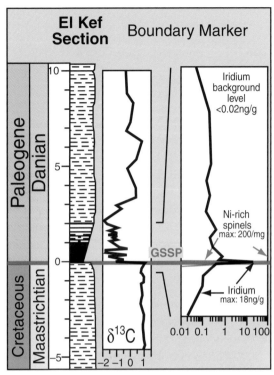

Figure 13.3. Stratigraphy of the base-Paleogene GSSP in the section at El Kef, Tunisia, with the primary boundary markers.

basal-Danian GSSP, and the base-Neogene GSSP was ratified in 1996.

The Paleogene Period begins with a global catastrophe – the famous bolide impact that dramatically terminated dinosaurs, ammonites, and the majority of other land animals and planktonic marine life on the planet. The GSSP is unique in its placement at the residual of that impact (clay layer with iridium anomaly, anomalous Ni-rich spinels, and mass extinction of microfossils) at El Kef in Tunisia. The basal Danian stage was originally part of the prior Cretaceous System, in part because the Danish type section was a continuation of the *creta* (= chalk) facies.

International subdivisions of Paleogene

The concepts of the Paleocene, Eocene and Oligocene series and their component stages has progressively mutated during the past century. In order to provide definitions that can be rigorously applied in both marine and terrestrial facies, the Paleogene Subcommission has tried to incorporate global markers, especially stable isotopic excursions and magnetic reversals, in the GSSP placements.

The middle Paleocene (Selandian Stage) begins with a major sea-level fall followed by a relative minimum in carbon-13 that is approximately 0.5 myr after the onset of Chron C26r. The top of Chron C26r is the main marker for the base of the upper Paleocene (Thanetian Stage). The Eocene Series (Ypresian Stage) begins with a pronounced negative excursion in carbon-13 that is associated with a dramatic warming episode ("thermal maximum") in the Earth's climate. The Oligocene Series begins with a local microfossil extinction, but is closely tied to a climatic cooling episode. Definitions of the stages within the Eocene and Oligocene will probably follow similar practices, but the GSSPs await international agreement.

Table 13.1 GSSPs of Paleogene stages, with location and primary correlation criteria (status in 2008)

Stage	GSSP location	Latitude, longitude	Boundary level	Correlation events	Reference
Chattian	*Possibly Monte Cagnero, Umbria-Marche region, Italy*			Potentially extinction of planktonic foraminifer *Chiloguembelina* (base Foram Zone P21b); or an isotopic/climatic event	
Rupelian	Massignano, near Ancona, Italy	43° 31′ 58.2″ N 13° 36′ 03.8″ E[a]	Base of a 0.5-m thick greenish-grey marl bed 19 m above base of section	Foraminifer LAD *Hantkenina* and *Cribrohantkenina*	*Episodes* **16**(3), 2001
Priabonian	*Alano section, Piave River; Veneto Prealps, Belluno Province, N Italy*		Tiziano Bed	Near FAD of calcareous nannofossil *Chiasmolithus oamaruensis* (base Zone NP18)	
Bartonian	*Contessa highway section near Gubio, Central Apennines, Italy*			Near extinction of calcareous nannofossil *Reticulofenestra reticulata*	
Lutetian	*Candidate is Agost section, Murcia Province, Betic Cordilleras, Spain*			Planktonic foraminifer FAD of *Hantkenina*	
Ypresian	Dababiya, near Luxor, Egypt	25° 30′ N 32° 31′ 52″ E[b]	Base of Bed 1 in DBH subsection	Base of the Carbon Isotope Excursion (CIE)	*Micropaleontology* **49**(1), 2003
Thanetian	*Zumaia section, N Spain*	43° 17.98′ N 2° 15.63′ W	30.5 m above the base of the Itzurun Formation	Magnetic polarity chronozone, base of C26n	
Selandian	*Zumaia section, N Spain*	43° 17.98′ N 2° 15.63′ W	Base of the red marls of Itzurun Formation	Onset of a sea-level drop and radiation of fasciculith group of calcareous nannofossils	
Danian *(base Paleogene)*	Oued Djerfane, west of El Kef, Tunisia	36° 09′ 13.2″ N 8° 38′ 54.8″ E	Reddish layer at the base of the 50-cm thick, dark boundary clay	Iridium geochemical anomaly. Associated with a major extinction horizon (foraminifers, calcareous nannofossils, dinosaurs, etc.)	*Episodes* **29**(4), 2006

a. According to Google Earth.
b. Derived from map.
Source: Details on each GSSP are available at *www.stratigraphy.org* and in the *Episodes* publications.

Selected aspects of Paleogene stratigraphy

Biostratigraphy

Planktonic foraminifers and calcareous nannofossils (especially coccolith plates of algae) have very refined and standardized zonations that are applicable in most ocean basins. Siliceous tests of radiolarians and diatoms, organic cysts of dinoflagellates, and different types of benthic foraminifers are other tools for correlating marine strata. The Paleogene was once called the *Nummulitique* period after the characteristic nummulites, which were large lenticular benthic foraminifers that occurred in rock-forming abundance in some tropical settings (e.g., the limestone used to build the Pyramids).

With the demise of dinosaurs, mammals expanded and rapidly evolved to fill terrestrial niches, then entered oceanic realms after the Paleocene. Intricate mammal zonations are available for each continent. Huge flightless birds, the diatrymas, were predators during the Paleocene and Eocene epochs.

Magnetic stratigraphy

The correlation of the C-sequence of marine magnetic anomalies to oceanic and terrestrial biostratigraphy is the main tool for assigning relative ages. In some intervals, such as the entire Selandian Stage and the 3 myr spanning the Paleocene–Eocene boundary, the Earth's polarity remained constant, but throughout most of the Paleogene, the resolution of magnetic polarity chrons is equivalent to microfossil zones.

Stable-isotope stratigraphy and dissolution events

The Paleogene underwent a complex history of warming and cooling. Ocean drilling has revealed the existence of several widespread dissolution levels during the Paleocene and Eocene that are interpreted as deep-sea responses to warming episodes (hyperthermal events), especially the major excursion at the Paleocene–Eocene boundary, that were superimposed on a long-term warming trend or sustained climatic optimum. Dissolution events that are used for interbasin correlation include the Danian–Selandian transition near the base of Chron C26r, the ELPE (Early Late Paleocene Event; also called the MPBE for Mid-Paleocene Biotic Event near base of Chron C26n), the PETM (Paleocene–Eocene Thermal Maximum at the base of Eocene), the ELMO (Early Eocene Layer of Mysterious Origin just below the top of Chron C24r), and the

X-event (near the base of Chron C24n.2n). The coincidence of carbon-12-enriched excursions (negative carbon-13 peaks) with some of these events has been interpreted as anomalous enrichments of the atmosphere–ocean system with methane and/or carbon dioxide.

The beginning of the Oligocene is synchronous with a surge in the oxygen-18 values in benthic foraminifers that was caused by a cooling of the deep ocean and ice accumulation over Antarctica (Oi-1 glaciation). Another glacial excursion (Mi-1) occurs at the end of the Oligocene.

Cycle and sequence stratigraphy

Ocean drilling cores and uplifted oceanic sediments are commonly characterized by oscillations in carbonate–clay content or other physical properties that were induced by orbital-climate effects on pelagic productivity and other marine processes. Analysis of the different Milankovitch cycle components has enabled high-resolution scaling for most of the Paleogene, and will soon connect to the astronomical time scale of the Neogene. Sea-level oscillations on longer time scales are one of the main controls on offshore petroleum reservoirs in clastic-margin settings.

Numerical time scale (GTS04, corrections, and future developments)

The numerical scaling of the Paleogene in GTS04 was constrained by a precise age of 65.5 Ma on the base-Cenozoic impact event according to an array of Ar–Ar radiometric dates. Durations of Paleocene magnetic polarity chrons from preliminary cycle stratigraphy relative to the base Cenozoic enabled assignment of a spreading-rate model to the Paleocene portion of the C-sequence, and the Eocene–Oligocene portion was calibrated by a spline-fit to applicable Ar–Ar radiometric dates. This C-sequence age model was used to assign ages to all other events.

Detailed cycle stratigraphy of several Paleogene sections has yielded an enhanced Paleocene through early Eocene scaling (e.g., Dinarès-Turell et al., 2007; Westerhold et al., 2008). This "floating" scale awaits a precise correlation of its 400-kyr long-eccentricity "master beat" to the progressive downward extension of the astronomical tuning of the Neogene. U–Pb age-dating of the base Cenozoic and Ar–Ar radiometric dating of astronomical-tuned ash beds have indicated that the base-Cenozoic boundary may be 0.5 to 1.0 myr older than the published age derived from previous Ar–Ar dating (Kuiper et al., 2008); thereby indicating that the Ar–Ar monitor "standard" and all Ar–Ar ages should be shifted to older values. As with the Mesozoic, the Paleogene scale in this book has retained the GTS04 age scale that was based on previously published values for Ar–Ar ages.

Base of Selandian is officially defined at a younger position (61.1 Ma)

In GTS04, the base of the yet-to-be-defined Selandian was provisionally assigned as the base of polarity Chron C26r, following suggestions by earlier workers. In the summer of 2007, the Paleocene working group decided to place the base of the international Selandian at the initiation of a sea-level drop (Exxon/Hardenbol sequence boundary "Se1") corresponding to the base of red marls in the Zumaia section of northern Spain. The best correlation criterion for the boundary is the diversification of the Fasciculith group of calcareous nannoplankton, an event preceding the lowest *Fasciculithus tympaniformis* which defines the base of nannoplankton zone NP5. Integrated magnetic and cycle stratigraphy indicates that this GSSP is approximately 0.65 myr (~33 precession cycles) above the base of Chron C26r. Therefore, maintaining the same C-sequence age scale as in GTS04 (base of Chron C26r = 61.7 Ma), the base of the Selandian is ~61.1 Ma, which is 0.6 myr younger than the working definition used in GTS04. This revision in the assigned level of the GSSP does not affect the relative ages of any biostratigraphic, magnetic, geochemical or other events. [Birger Schmitz, chair of Paleocene working group, contributed to this revision.]

Acknowledgements

For further details/information, we recommend "The Paleogene Period" by H. P. Luterbacher et al. (in *A Geologic Time Scale 2004*). Portions of the background material are from documents of the Paleogene Subcommission.

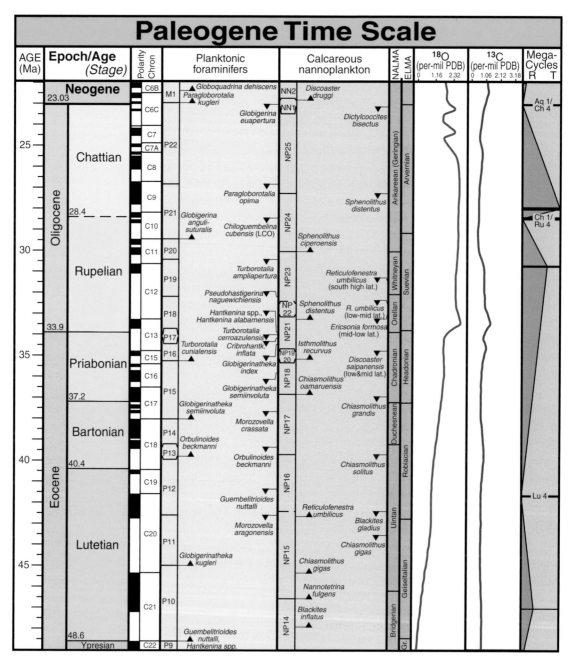

Figure 13.4. Numerical ages of epoch/series and age/stage boundaries of the Paleogene with major marine biostratigraphic zonations and principle eustatic trends. ["Age" is the term for the time equivalent of the rock-record "stage".] The planktonic foraminifer scale is modified from Berggren *et al.* (1995a, 1995b) with age updates from Berggren and Pearson (2005). The calcareous nannoplankton scale is modified from tables in Berggren *et al.* (1995a, b) and from explanatory notes of ODP Initial Reports. Mammal stage from North America (NALMA) and Europe (ELMA) are from Hooker (in Paleogene chapter of GTS04). The ^{13}C and ^{18}O curves are generalized from Zachos *et al.* (2001). The Mega cycles are from Hardenbol *et al.* (1998).

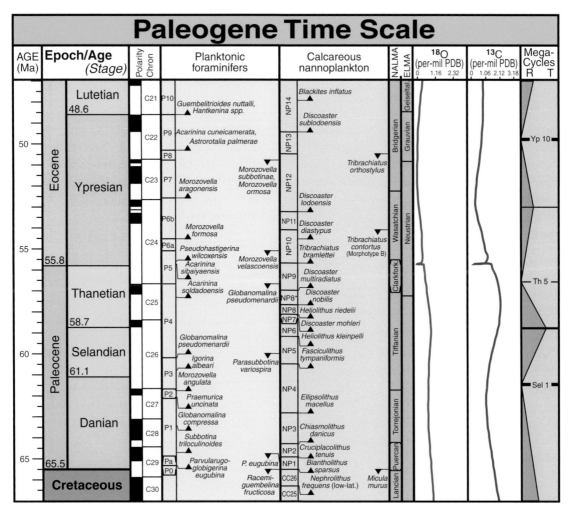

Figure 13.4. (cont.)

Further reading

Berggren, W. A., and Pearson, P. N., 2005. A revised tropical to subtropical Paleogene planktonic foraminiferal zonation. *Journal of Foraminiferal Research*, **35**: 279–298.

Berggren, W. A., Kent, D. V., Swisher, C. C.III, and Aubry, M. -P., 1995a. A revised Cenozoic geochronology and chronostratigraphy. In: *Geochronology Time Scales and Global Stratigraphic Correlation*, eds. W. A. Berggren *et al. Society of Economic Paleontologists and Mineralogists Special Publication*, **54**: 129–212.

Berggren, W. A., Hilgen, F. J., Langereis, C. G., Kent, D. V., Obradovitch, J. D., Raffi, I., Raymo, M., and Shackleton, N. J., 1995b. Late Neogene (Pliocene–Pleistocene) chronology: new perspectives in high-resolution stratigraphy. *Geological Society of America Bulletin*, **107**: 1272–1287.

Bernaola, G., Baceta, J. I., Payros, A., Orue-Etxebarria, X., and Apellaniz, E. (eds.), 2006. *Climate and Biota of the Early Paleogene 2006: Post-Conference Field Excursion Guidebook – Zumaia Section*. Bilbao: Department of Stratigraphy and Paleontology, University of the Basque Country.

de Graciansky, P.-C., Hardenbol, J., Jacquin, Th., and Vail, P. R (eds.), 1998. *Mesozoic–Cenozoic Sequence Stratigraphy of European Basins. SEPM Special Publication* **60**.

Dinarès-Turell, J., Baceta, J. I., Bernaola, G., Orue-Etxebarria, and Pujalte, V., 2007. Closing the Mid-Paleocene gap: toward a complete astronomically tuned Paleocene Epoch and prospective Selandian and Thanetian GSSPs at Zumaia (Basque Basin, W. Pyrenees). *Earth and Planetary Science Letters*, **262**: 450–467.

Hardenbol, J., Thierry, J., Farley, M. B., Jacquin, Th., de Graciansky, P.-C., and Vail, P. R. (with numerous contributors), 1998. Mesozoic and Cenozoic sequence Chronostratigraphic framework of European basins. In: *Mesozoic–Cenozoic Sequence Stratigraphy of European Basins*, eds. P.-C. de Graciansky, J. Hardenbol, Th. Jacquin, and P. R. Vail. *SEPM Special Publication*, **60**: 3–13, 763–781, and chart supplements.

Kroon, D., Zachos, J. C., and ODP Leg 208 Scientific Party, 2007. Leg 208 synthesis: Cenozoic climate cycles and excursions. In: *Early Cenozoic Extreme Climates: The Walvis Ridge Transect*, eds. D. Kroon, J. C. Zachos, and C. Richter. *Proceedings of the Ocean Drilling Program, Scientific Results* **208**: 1–55. Available on-line at www-odp.tamu.edu/publications/208_SR/synth/synth.htm

Kuiper, K. F., Deino, A., Hilgen, F. J., Krijgsman, W., Renne, P. R., and Wijbrans, J. R., 2008. Synchronizing the rock clocks of Earth history. *Science*, **320**: 500–504.

Molina, E., Alegret, L., Arenillas, I., Arz, J. A., Gallala, N., Hardenbol, J., von Salis, K., Steuraut, E., Vandenberge, N., and Zaghib-Turki, D., 2006. The Global Boundary Stratotype Section and Point for the base of the Danian Stage (Paleocene, Paleogene, "Tertiary", Cenozoic) at El Kef, Tunisia: original definition and revision. *Episodes*, **29**(4): 263–278.

Prothero, D. R., 1994. *The Eocene–Oligocene Transition: Paradise Lost*. New York: Cambridge University Press.

Westerhold, T., Röhl, U., Raffi, I., Fornaciari, E., Monechi, S., Reale, V., Bowles, J., and Evans, H. F., 2008. Astronomical calibration of the Paleocene time. *Palaeogeography, Palaeoclimatology, Palaeoecology*, **257**: 377–403.

Wing, S. L., Gingerich, P. D., Schmitz, B., and Thomas, E. (eds.), 2003. *Causes and Consequences of Globally Warm Climates in the Early Paleogene. Geological Society of America Special Paper* **369**.

Zachos, J. C., Pagani, M., Sloan, L., Thomas, E., and Billups, K., 2001. Trends, rhythms, and aberrations in global climate 65 Ma to present. *Science*, **292**: 686–693.

Selected on-line references

Paleogene Subcommission – *wzar.unizar.es/ isps/index.htm* – GSSP information and working group; PDFs of papers relevant to Paleogene.

Paleocene Mammals of the World (compiled by Martin Jehle, 2006) – *www.paleocene-mammals. de* – especially the "Introduction to Paleocene mammals" ["*superb coverage of Paleocene Mammals – Best on the Web*" – Palaeos link]

We recommend the extensive Paleogene webpages and links at *Palaeos*, Smithsonian Institution, University of California Museum of Paleontology, and *Wikipedia*. See URL details at end of Chapter 1.

14
Neogene Period

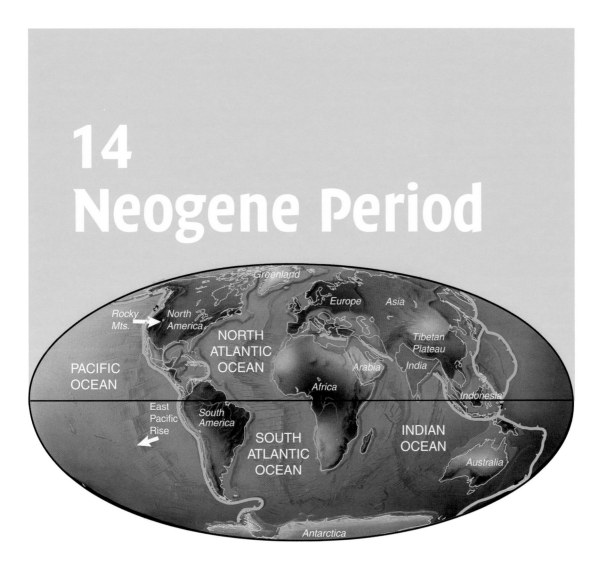

Figure 14.1. Geographic distribution of the continents during the Neogene Period (14 Ma). The paleogeographic map was provided by Christopher Scotese.

History and base of Neogene

The original "Neogen ("new", "clan/birth") Stufe" of Moritz Hörnes was introduced in 1853/1864 to differentiate the younger molluscan fauna of the Vienna Basin from those of the Eocene (sensu Lyell, 1833). According to this division of the Molasse Group, the Neogene strata also included the *Knochen-Höhlen und der Löss* or glacial-derived deposits that are typical of "Quaternary,"

although the historical record is ambiguous (see discussions in Neogene chapter of GTS04; Walsh, 2008; and Lourens, 2008). Even though usage of "Neogene" by marine stratigraphers customarily included the full suite of Miocene through Holocene epochs, many continental workers preferred to distinguish

Figure 14.2. The GSSP marking the base of the Neogene System and its lowermost Aquitanian Stage at Lemme-Carrosio, Italy. GSSP level is 35 m below the top of the light-gray clayey silts (Rigoroso Formation), approximately where the exposed barren ridge on the right descends and becomes obscured by foreground vegetation. It is near a thin relatively indurated lighter-colored band. Photograph from the Neogene Subcommission website, *www.geo.uu.nl/sns.*

the glacial-climate-dominated uppermost Cenozoic as a separate Quaternary Period/ System. In 2006–2007, the IUGS requested ICS in collaboration with the International Union for Quaternary Research (INQUA) to establish a Quaternary Period (see chapter on Quaternary). However, for convenience, we will summarize the oceanic stratigraphy and astronomical tuning of the time scale for the past 23 myr (Miocene through Holocene) in this Neogene chapter, and the continental records of the past 2.5 myr in the Quaternary chapter.

The base of the Neogene System, the Miocene Series, and the Aquitanian Stage is at the base of polarity Chron C6Cn.2n. This level corresponds approximately with oxygen-isotope positive (cooling) event Mi-1 and the associated major sequence boundary "Aq1," and coincides with several microfossil events. The cooling event was caused by an infrequent superimposition of different orbital cycles (Milankovitch cycles) that led to a prolonged interval lacking high-amplitude

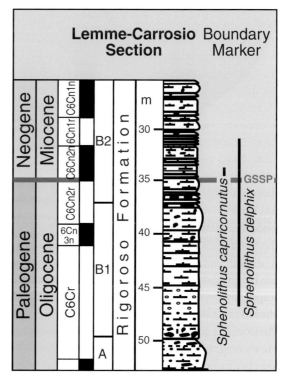

Figure 14.3. Stratigraphy of the base-Neogene GSSP in the section at Lemme-Carrosio, Italy, with the primary boundary markers.

insolation variations on Milankovitch time scales; therefore, it can be precisely dated as 23.03 Ma.

International subdivisions of Neogene

The progressively shorter duration of the four epochs (Miocene, Pliocene, Pleistocene and Holocene) of late Cenozoic reflects the relative importance and preservation of deposits as one approaches the present day. In order to facilitate high-resolution dating of events, the Neogene Subcommission selected the sections that contain the GSSPs defining these series and their component stages to have a record of sedimentary cycles produced by orbital-climate oscillations. Therefore, the biostratigraphic, magnetic and geochemical events that are associated with the GSSP levels are directly correlated to the astronomical-tuned time scale. As of Spring 2008, this combined GSSP and astronomical scaling has been extended back to the Serravallian GSSP (13.82 Ma). The primary marker for the Serravallian GSSP is a cooling event reflected in oxygen isotopes (event Mi3b). Microfossils are the primary markers for the upper Miocene stages, and magnetic polarity reversals were chosen for placement of most Pliocene and Pleistocene stage GSSPs. In order to establish cycle-constrained placement of GSSPs for the lower Miocene stages of Langhian and Burdigalian, it may be necessary to designate GSSPs in ocean drilling cores with an auxiliary GSSP in a less-cyclic-exposed outcrop.

Selected aspects of Neogene stratigraphy

Biostratigraphy

Mammal remains are the most useful fossils for correlating Cenozoic terrestrial sediments. Shallow-marine and brackish deposits yield regional mollusc and larger foraminiferal zonations. In oceanic settings, the array of microfossils include pelagic and benthic foraminifers, calcareous algae, diatoms, radiolarians and dinoflagellate cysts, but calcareous plankton (planktonic foraminifera and calcareous nannofossils) are most important.

Magnetic stratigraphy

The C-sequence of marine magnetic anomalies has been correlated via magnetostratigraphy to oceanic and terrestrial deposits. The relatively high frequency of magnetic reversals, particularly in the Miocene, generally requires constraints from biostratigraphy in the strata for unambiguous polarity chron assignments. In general, each microfossil zone can be subdivided by at least two polarity chrons, thereby enabling a very high-resolution integrated scale.

Sequence stratigraphy

Neogene sea-level oscillations are a major control on the deposit of shallow-marine sequences and of deep-water sands off clastic margins. In the Gulf of Mexico and other basins with high sedimentation rates, these lowstand sands are the main offshore petroleum

Table 14.1 *GSSPs of Neogene stages, with location and primary correlation criteria (status in 2008). The Gelasian is currently in the Pliocene, but a pending proposal to define the Quaternary would shift it to the Pleistocene.*

Stage	GSSP location	Latitude, longitude	Boundary level	Correlation events	Reference
Gelasian *(currently in Pliocene)*	Monte San Nicola, Sicily, Italy	37° 08′ 48.8″ N 14° 12′ 12.6″ E[a]	Base of marly layer overlying sapropel MPRS 250 with an age of 2.588 Ma	Precessional cycle 250 from the present, Marine Isotope Stage 103, with an age of 2.588 Ma	*Episodes* **21**(2), 1998
Piacenzian	Punta Piccola, Sicily, Italy	37° 17′ 20″ N 13° 29′ 36″ E[a]	Base of the beige marl bed of small-scale carbonate cycle 77 with an age of 3.6 Ma	Precessional excursion 347 from the present with an astrochronological age estimate of 3.6 Ma	*Episodes* **21**(2), 1998
Zanclean	Eraclea Minoa, Sicily, Italy	37° 23′ 30″ N 13° 16′ 50″ E	Base of the Trubi Formation	Insolation cycle 510 counted from the present with an age of 5.33 Ma	*Episodes* **23**(3), 2000
Messinian	Oued Akrech, Morocco	33° 56′ 13″ N 6° 48′ 45″ W	Reddish layer of sedimentary cycle number 15	First regular occurrence of planktonic foraminifer *Globorotalia miotumida* and the FAD of the calcareous nannofossil *Amaurolithus delicatus*	*Episodes* **23**(3), 2000
Tortonian	Monte dei Corvi Beach, near Ancona, Italy	43° 35′ 12″ N 13° 34′ 10″ E	Mid-point of sapropel layer of basic cycle number 76	Last Common Occurrences of the calcareous nannofossil *Discoaster kugleri* and the planktonic foraminifer *Globigerinoides subquadratus*	*Episodes* **28**(1), 2005
Serravallian	Ras il Pellegrin section, Fomm Ir-Rih Bay, W coast of Malta	35° 54′ 50″ N 14° 20′ 10″ E	Formation boundary between the Globigerina Limestone and Blue Clay	Mi3b oxygen-isotopic event (global cooling episode); near LAD of calcareous nannofossil *Sphenolithus heteromorphus*	
Langhian	Potentially in astronomically tuned ODP core (Leg 154) or in Italy (Moria or La Vedova)			Near FAD of planktonic foraminifer *Praeorbulina glomerosa* and top of magnetic polarity chronozone C5Cn.1n	
Burdigalian	Potentially in astronomically tuned ODP core			Near FAD of planktonic foraminifer *Globigerinoides altiaperturus* or near top of magnetic polarity chronozone C6An	
Aquitanian *(base Neogene)*	Lemme-Carrosio Section, Allessandria Province, Italy	44° 39′ 32″ N 08° 50′ 11″ E	35 m from the top of the section	Base of magnetic polarity chronozone C6Cn.2n; FAD of planktonic foraminifer *Paragloborotalia kugleri*; near extinction of calcareous nannofossil *Reticulofenestra bisecta* (base Zone NN1); oxygen-isotopic event Mi-1	*Episodes* **20**(1), 1997

a. According to Google Earth.
Source: Details on each GSSP are available at *www.stratigraphy.org* and in the *Episodes* publications.

reservoirs, therefore have been carefully correlated to other biostratigraphic events.

Cycle and stable-isotope stratigraphy

Neogene sediments recovered by ocean drilling cores or uplifted in tectonic regions, such as the southern Italian margin, are commonly characterized by high-frequency variations in oxygen-18, sediment composition and/or physical properties. In addition to unlocking past changes in the state of the ocean-climate system, the oxygen-18 variations are extremely useful for chronostratigraphic correlation, and have a systematic numbering system ("marine isotope stages" or MIS) that extends from the Present into the upper Miocene. In the Plio-Pleistocene, the episodes of enriched oxygen-18 (cold or glacial intervals) are given even numbers. The major MIS 110 near the base of the Gelasian Stage corresponds to major sequence boundary "Ge1" and is correlated to the oldest glacial "Quaternary" deposits that extended over a significant portion of the northern continents. These oscillations in stable isotopes and sediment characteristics were produced by Milankovitch orbital-climate cycles, thereby enabling compilation of an astronomical time scale.

Carbon-13 reveals distinct variability on Milankovitch time scales. On longer time scales, the Miocene records two maxima in carbon-13. The Paleogene–Neogene transition is within the first peak, and the second "Monterey Event" coincides approximately with a mid-Miocene climatic optimum. The late Miocene shift toward lighter carbon-13 values in marine sediments is partially due to the expansion of plants that use the C_4 pathway of photosynthesis (especially grasses) and partially offsetting a general shift toward heavier values in the continental record.

Numerical time scale (GTS04, corrections, and future developments)

The astronomical time scale for the cyclic deposits of Neogene has attained an extraordinary level of precision. A refined astronomical solution for the Solar System variables that control these Milankovitch cycles was used in *A Geologic Time Scale 2004* to assign ages to within 1000 years to each cycle. Each GSSP that has been established for the Miocene through Pleistocene has been assigned in cyclic facies with a known correlation to the master orbital-climate cycle scale, thereby enabling its age to be assigned with a precision of less than 10 000 years.

Levels of first and last pelagic foraminifer and calcareous nannoplankton occurrences have been independently calibrated in each ocean basin to the orbital-cycle scale, thereby indicating the degree of interbasin diachroneity.

The superposition of longer-term eccentricity cycles and modulation of obliquity produces periodic "nodes." The node associated with the climatic change at the

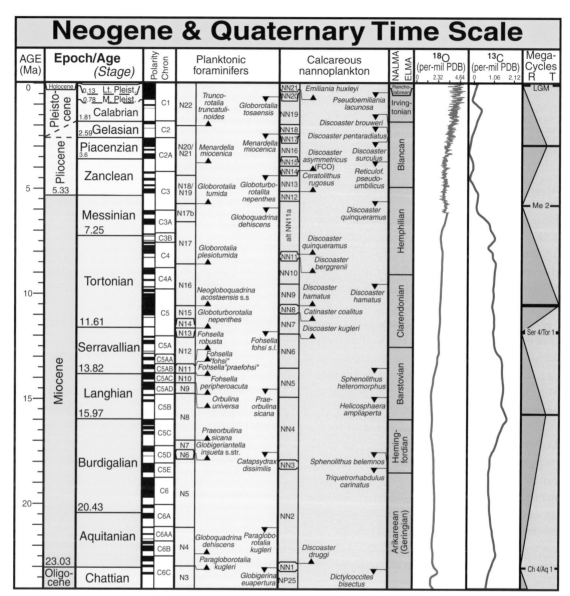

Figure 14.4. Numerical ages of epoch/series and age/stage boundaries of the Neogene–Quaternary with major marine biostratigraphic zonations and principal eustatic trends. ["Age" is the term for the time equivalent of the rock-record "stage".] The planktonic foraminifer and calcareous nannofossil scales are from Berggren *et al.* (1995a, b), with revised age calibrations from Lourens *et al.* (GTS04 appendix). Mammal stages from North America (NALMA) and Europe (ELMA) are from Hooker (in Paleogene chapter of GTS04). The ^{13}C and ^{18}O curves are generalized from Zachos *et al.* (2001), with high-resolution ^{18}O curve for the past 6 myr from Crowhurst (2002). The Mega cycles are from Hardenbol *et al.* (1998). Definition of the Quaternary and revision of the Pleistocene are under discussion. Base of the Pleistocene is at 1.81 Ma (base of Calabrian), but may be extended to 2.59 Ma (base of Gelasiàn). The historic "Tertiary" comprises the Paleogene and Neogene, and has no official rank.

Oligocene–Miocene boundary has an astronomical-derived age of 23.03 Ma.

In *A Geologic Time Scale 2004*, events within the interval between approximately 14 Ma (= oldest cycle with direct correlation to biostratigraphy and magnetic chrons) and 23 Ma (= the long-cycle-calibrated Oligocene–Miocene boundary) were assigned ages according to their magnetostratigraphic placement with respect to a smoothed spreading-rate model for that portion of the C-sequence marine magnetic anomalies. In the meantime, with the progressive extension of the bio-astronomical scale into older Miocene, and eventually Paleogene, oceanic deposits, this lower and middle Miocene interval already has its own suite of precise orbital age assignments.

Miocene stages of Burdigalian through Serravallian were awaiting formal GSSP definitions in GTS04, therefore their ages were provisionally assigned according to potential microfossil-based markers. However, in some cases, the eventual international decision on defining these stages may result in a different primary marker. The GSSP ratified in 2006 for the Serravallian Stage is one example.

Base of Serravallian is officially defined at an older (~0.2 myr) position

In GTS04, the base of the Serravallian was placed at the last occurrence of calcareous nannofossil *Sphenolithus heteromorphus*, which has an astronomical-tuned age of 13.654 Ma in the Mediterranean. The Neogene Subcommission proposed in 2006 to place the base-Serravallian GSSP in accordance with the end of the major Mi-3b cooling step in oxygen isotopes, which reflects a major increase in Antarctic ice volume marking the Earth's final transition into "Icehouse" climate. The GSSP was ratified in late 2006. This GSSP level has been astronomically dated as 13.82 Ma. This official decision in the level of the GSSP does not affect the ages of any biostratigraphic, magnetic, geochemical or other events in GTS04, but is merely the placement of the stage-boundary definition relative to those events.

Acknowledgements

Frits Hilgen (chair of Neogene Subcommission) contributed extensively to this overview, and Lucas Lourens provided a copy of his in-press review of Neogene. For further details/information, we recommend "The Neogene Period" by L. Lourens, F. Hilgen, N. J., Shackleton, J. Laskar, and D. Wilson (in *A Geologic Time Scale 2004*). Portions of the background material are from documents of the Neogene Subcommission.

Further reading

Abels, H. A., Hilgen, F. J., Krijgsman, W., Kruk, R. W., Raffi, I., Turco, E., and Zachariasse, W. J., 2005. Long-period orbital control on middle Miocene global cooling: integrated stratigraphy and astronomical tuning of the Blue Clay Formation on Malta. *Paleoceanography*, **20**: PA4012, doi: 10.1029/2004 PA001129.

Berggren, W. A., 2007. Status of the hierarchical subdivision of higher order marine Cenozoic chronostratigraphic units. *Stratigraphy*, **4**: 99–108.

Berggren, W. A., Kent, D. V., Swisher, C. C.III, and Aubry, M.-P., 1995a. A revised Cenozoic geochronology and chronostratigraphy. In: *Geochronology Time Scales and Global Stratigraphic Correlation*, eds. W. A. Berggren *et al. Society of Economic Paleontologists and Mineralogists Special Publication*, **54**: 129–212.

Berggren, W. A., Hilgen, F. J., Langereis, C. G., Kent, D. V., Obradovitch, J. D., Raffi, I., Raymo, M., and Shackleton, N. J., 1995b. Late Neogene (Pliocene–Pleistocene) chronology: new perspectives in high-resolution stratigraphy. *Geological Society of America Bulletin*, **107**: 1272–1287.

Crowhurst, S. J., 2002. Composite isotope sequence. *The Delphi Project*. Available on-line at www.esc.cam.ac.uk/new/v10/research/institutes/godwin/body.html.

de Graciansky, P.-C., Hardenbol, J., Jacquin, Th., and Vail, P. R (eds.), 1998. *Mesozoic–Cenozoic Sequence Stratigraphy of European Basins. SEPM Special Publication* 60.

Hardenbol, J., Thierry, J., Farley, M. B., Jacquin, Th., de Graciansky, P.-C., and Vail, P. R. (with numerous contributors), 1998. Mesozoic and Cenozoic sequence Chronostratigraphic framework of European basins. In: *Mesozoic–Cenozoic Sequence Stratigraphy of European Basins*, eds. P.-C. de Graciansky, J. Hardenbol, Th. Jacquin, and P. R. Vail. *SEPM Special Publication*, **60**: 3–13, 763–781, and chart supplements.

Hilgen, F. J., Brinkhuis, H., and Zachariasse, W. J., 2006. Unit stratotypes for global stages: the Neogene perspective. *Earth-Science Reviews*, **74**: 113–125.

Hilgen, F. J., Kuiper, K., Krijgsman, W., Snel, E., and van der Laan, E., 2007. Astronomical tuning as the basis for high resolution chronostratigraphy: the intricate history of the Messinian Salinity Crisis. *Stratigraphy*, **4**: 151–158.

Holbourn, A., Kuhnt, W., Schulz, M., and Erlenkeuser, H., 2005. Impacts of orbital forcing and atmospheric carbon dioxide on Miocene ice-sheet expansion. *Nature*, **438**: 483–487.

Holbourn, A., Kuhnt, W., Schulz, M., Flores, J.-A. and Andersen, N., 2007. Orbitally-paced climate evolution during the middle Miocene "Monterey" carbon-isotope excursion. *Earth and Planetary Science Letters*, **261**: 534–550.

Lisiecki, L. E., and Raymo, M. E., 2005. A Pliocene–Pleistocene stack of 57 globally distributed benthic $\delta^{18}O$ records. *Paleoceanography*, **20**: PA1003, doi:10.1029/2004 PA001071.

Lourens, L. J., 2008. On the Neogene–Quaternary debate. *Episodes*, in press.

Lyell, C., 1833. *Principles of Geology*, vol. 3. London: John Murray.

Raffi, I., Backman, J., Fornaciari, E., Pälike, H., Rio, D., Lourens, L., and Hilgen, F., 2006. A review of calcareous nannofossil astrobiochronology encompassing the past 25 million years. *Quaternary Science Reviews*, **25**: 3113–3137.

Van Dam, J. A., Abdul Aziz, H. A., Sierra, M. A. A., Hilgen, F. J., van den Hoek Oostende, L. W., Lourens, L. J., Mein, P., van der Meulen, A. J., and Pelaez-Campomanes, P. 2006. Long-period astronomical forcing of mammal turnover. *Nature*, **443**: 687–691.

Walsh, S. L., 2008. The Neogene: origin, adoption, evolution, and controversy. *Earth Science Reviews*, **89**: 42–72.

Zachos, J., Shackleton, N. J., Revenaugh, J. S., Pälike, H., and Flower, B. P., 2001. Climate response to orbital forcing across the Oligocene–Miocene boundary. *Science*, **292**: 274–278.

Selected on-line references

Neogene Subcommission – *www.geo.uu.nl/sns* – details of GSSPs, including PDFs.

Messinian online "living in an evaporitic world" – *www.messinianonline.it* – exploration of the great environmental changes experienced by the whole Mediterranean area when that sea dried at about 6 millions years ago.

We recommend the extensive Neogene webpages and links at *Palaeos*, Smithsonian Institution, University of California Museum of Paleontology, and *Wikipedia*. See URL details at end of Chapter 1.

15 Quaternary Period

Philip Gibbard, Kim Cohen, and James Ogg

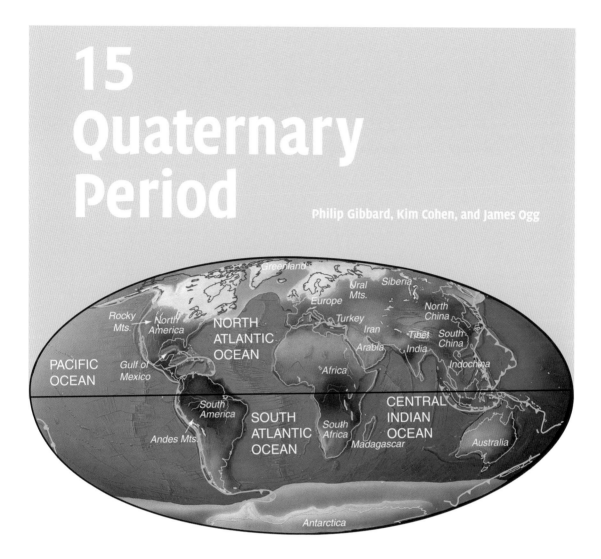

History and base of Quaternary

Figure 15.1. Geographic distribution of the continents during the Quaternary Period (18 000 yrs). The paleogeographic map was provided by Christopher Scotese.

Despite being the most widely used unit in field mapping and having the greatest number of active researchers, the interval known as *Quaternary* is unique among the chronostratigraphic subdivisions in having the most controversial definition and rank. The convoluted history and divergent concepts of Quaternary usage is fraught with opinionated debate, beginning with the early International Geological Congresses which considered relegating *Quaternary* to be an un-ranked synonym for a vaguely defined *Pleistocene* epoch (1894) or "Modern" period

(1900). The association of *Quaternary* with the "Ice Ages" created another problem after the investigation of new regions, improved dating methods and deep-sea oxygen isotope records. The onset of these continental glaciations was discovered to begin much earlier in the Neogene than the ratified base of the Pleistocene Series.

In 1983, the base-Pleistocene GSSP was ratified at Vrica, Italy, near the top of the Olduvai magnetic Subchron (~1.8 Ma), but

Table 15.1 GSSPs of Quaternary stages, with location and primary correlation criteria (status in 2008). Stage terminology for the upper three stages has not yet been formalized. The Gelasian is currently in the Pliocene, but a pending proposal to define the Quaternary would shift it to the Pleistocene

Stage	GSSP location	Latitude, longitude	Boundary level	Correlation events	Reference
Holocene Series	NorthGRIP ice core, central Greenland	75.10° N 42.32° W	1492.45 m depth in Borehole NGRIP2	End of the Younger Dryas cold spell, which is reflected in a shift in deuterium excess values	Episodes, 2008 (in press)
Upper Pleistocene (Tarantian)	Amsterdam-Terminal borehole, Netherlands	52° 22′ 45″ N 4° 54′ 52″ E	63.5 m below surface	Base of warm marine isotope stage 5e, before final glacial episode of Pleistocene	Episodes, 2008 (in press)
Middle Pleistocene (Ionian)	Candidate sections in Italy (Montalbano Jorica or Valle di Manche) and Japan (Chiba)			Magnetic – Brunhes-Matuyama magnetic reversal (base of Chron 1n)	Episodes, 2008 (in press)
Lower Pleistocene (Calabrian)	Vrica, Italy	39° 02′ 18.61″ N 17° 08′ 05.79″ E	Base of the marine claystone overlying the sapropelic marker Bed 'e'	~15 kyr after end of Olduvai normal polarity chron	Episodes 8(2), 1985
Gelasian (currently in Pliocene)	Monte San Nicola, Sicily, Italy	37° 08′ 48.8″ N 14° 12′ 12.6″ E[a]	Base of marly layer overlying sapropel MPRS 250 with an age of 2.588 Ma	Precessional cycle 250 from the present, Marine Isotope Stage 103, with an age of 2.588 Ma	Episodes 21(2), 1998

a. According to Google Earth.
Source: Details on each GSSP are available at *www.stratigraphy.org* and in the *Episodes* publications.

the decision "*was isolated from other more or less related problems, such as ... status of the Quaternary*." The Gelasian Stage was later created (1996) to fill the "gap" between this GSSP and the "traditional" span of the Piacenzian Stage of the Pliocene Series. Unfortunately, neither ICS nor IUGS voted officially to clarify the definition and status of the Quaternary, although the base-Pleistocene GSSP was re-ratified by IUGS in 1998 following inconclusive discussions to revise it. Nevertheless, the Quaternary is commonly shown on time scale charts as a period/system subdivided into the Pleistocene and Holocene epoch/series.

The International Union of Quaternary Research (INQUA; under the International Council for Science) and its component national members unanimously voted (August, 2007)

that the "Quaternary Period spans the last 2.6 million years of Earth's history." This definition of the Quaternary is based on recognition of glacial-driven major oxygen-isotope excursions and pronounced eustatic lowstands on continental shelves, plus the improved dating of the onset of the main loess deposition in China, the earliest till deposits in Missouri, and other traditional "Quaternary" deposits. There is a dramatic change in deep-sea circulation patterns and ice-rafted debris into the northern oceans that corresponds to the earliest pronounced glacial interval (Marine Isotope Stage 110, with an astronomical-tuned age of 2.73 Ma) and a major eustatic lowstand (sequence boundary "Ge1"). The base of the Gelasian Stage is slightly younger (warm interval MIS 103; age of 2.59 Ma), but its association

Figure 15.2. The GSSP for the base of the Calabrian Stage (current base of Pleistocene Series) at Vrica, Italy. The sapropel marker beds b, c, d, and e in this lowermost part of section Vrica B are indicated. Photograph from the Neogene Subcommission website, *www.geo.uu.nl/sns*.

with the magnetic reversal at the onset of the Matuyama reversed-polarity Chron enables an unambiguous and precise global marker; therefore, the Quaternary was recommended by INQUA to be defined with this established Gelasian GSSP. In 2006–2007, the IUGS urged ICS that "it is necessary to reach as soon as possible an international consensus on the Quaternary problem that has to be ratified during the 2008 IGC," that the placement/rank of Quaternary should not violate the hierarchy of geological units, and this consensus should involve INQUA's opinion.

Therefore, in order to rectify the offset of Quaternary (as preferred by INQUA and many national usages) and the current GSSP of the Pleistocene Epoch/Series, the ICS and INQUA proposed (May, 2007) that the Gelasian Stage should be transferred to the Pleistocene, thereby enabling a Quaternary Period/System to be formally established within the Cenozoic. This is in accord with the 1948 unanimous decision by the International Geological Congress Council that the Pleistocene should include the Calabrian marine stage and the *Villafranchian* regional continental stage (which is now known to encompass the Gelasian

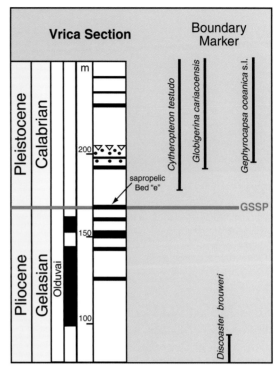

Figure 15.3. Stratigraphy of the Calabrian GSSP (current base of Pleistocene Series) in the section at Vrica, Italy, with the primary boundary markers.

Stage). The preceding period/system would remain the Neogene. This proposal was deferred by IUGS for further discussion at the August 2008 International Geological Congress (IGC). Therefore, the Quaternary currently (as of 2008) remains officially undefined. We display both possible definitions in the graphics – base of proposed Calabrian Stage (present base of Pleistocene Series; ~1.8 Ma) and base of Gelasian Stage (~2.6 Ma; as recommended by INQUA).

International subdivisions of Quaternary

The boundary between the Pleistocene and Holocene epochs/series is at 11 700 years before

Figure 15.4. The GSSP for the base of the Gelasian Stage (base of the Quaternary System as recommended by INQUA, and submitted by ICS to IUGS) at Lemme-Carrosio, Italy. Photograph from the Neogene Subcommission website, *www.geo.uu.nl/sns.*

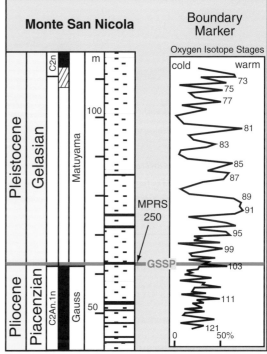

Figure 15.5. Stratigraphy of the Gelasian GSSP (base of the Quaternary System as recommended by INQUA, and submitted by ICS to IUGS) in the section at Lemme-Carrosio, Italy, with the primary boundary markers.

AD 2000. The base Holocene is assigned at the first signs of climatic warming at the end of the Younger Dryas/Greenland Stadial 1 cold phase, which is most clearly reflected in a shift in deuterium excess values, accompanied by changes in $\delta^{18}O$, dust concentration, a range of chemical species, and by a change in annual layer thickness in the NorthGRIP Greenland ice core (Walker *et al.*, 2008).

The current Pleistocene Series is proposed to be divided into three stages – a Calabrian Stage with its base at the current base-Pleistocene GSSP at Vrica, Italy, an Ionian Stage with its base at the onset of the Brunhes normal-polarity chron at 773 ka, and a Tarantian Stage that would span the last interglacial and glacial intervals (Marine Isotope Stages 5 to 2; beginning at about 125 ka). The Tarantian GSSP is 63.5 m below the Amsterdam Terminal for ocean liners, which was built on a lacustrine succession with an expanded record of the last interglacial, and has an auxiliary reference section at the corresponding *Eemian* regional stage stratotype. As indicated above, the Gelasian Stage may potentially become the

lowest stage in a Quaternary System and the associated extension of the Pleistocene Series. The stages will also be grouped into subseries divisions which are widely applied to Pleistocene chronological sequences. The Lower (Early) Pleistocene will comprise the Gelasian and Calabrian, the Middle Pleistocene the Ionian Stage, and the Upper (Late) Pleistocene will be equivalent to the Tarantian Stage. All of these stages and subseries await ratification.

The effects of humans during the past three centuries have dominated the distribution of terrestrial ecosystems, patterns of sediment accumulation, atmospheric concentrations of greenhouse gases, and even oceanic chemistry.

"The Anthropocene could be said to have started in the late eighteenth century, when analyses of air trapped in polar ice showed the beginning of growing global concentrations of carbon dioxide and methane" (Crutzen, 2002). Proposals to define formally an *Anthropocene* epoch or stage (e.g., Zalasiewicz *et al.*, 2008) are currently being discusssed.

Selected aspects of Quaternary stratigraphy

Stable-isotope stratigraphy

Marine isotope stages (MIS) based on oscillations in oxygen-18 values of benthic foraminifera enable a systematic ultra-high resolution of the Quaternary and the Pliocene. The even-numbered peaks (elevated values) in oxygen-18 are cold or glacial intervals, and odd-numbered troughs are warm or interglacial intervals. The Quaternary spans over 50 couplets of warm–cold climate; and most of these marine-isotope stages are recognized in patterns of loess accumulation, ice-core records, organic-rich bands in Mediterranean sediments (*sapropels*) and other climatic-sensitive strata.

Drilling into the Greenland ice cap and coring of North Atlantic sediments revealed that the last glacial-dominated period (MIS 4 to MIS 2; 110 to 11 ka) was not an extended structureless, cold interval, but experienced frequent climatic excursions. Over 20 brief warming events (Dansgaard–Oeschger cycles) are identified in the Greenland ice cores. Some of these were preceded by anomalous release of icebergs into the North Atlantic, as recognized by horizons of ice-rafted debris (Heinrich Events). Seven main Heinrich Events have been identified within the past 60 kyr.

Magnetic stratigraphy and biostratigraphy

Quaternary magnetic polarity intervals corresponding to marine magnetic anomalies C2r to C1n are commonly denoted by names of pioneers in geomagnetism (main chrons of Gauss, Matuyama and Brunhes) and type-localities for brief subchrons (e.g., Réunion, Olduvai, Jaramillo). Very brief reversed-polarity excursions within the Brunhes Chron have also been identified (e.g., Emperor at ~40 ka; Blake at ~12 ka), but are generally too fleeting to serve as useful magnetostratigraphic markers.

Terrestrial strata are commonly divided based upon characteristic mammalian fossil assemblages. Other techniques, such as pollen assemblages, have also been widely used in Eurasia. The teeth of voles, a mouse-like rodent, are particularly useful for subdivisions in Eurasia and North America. Biostratigraphical methods for marine strata are summarized in the Neogene chapter.

The Quaternary is the age of the genus *Homo*, with *Homo habilis* evolving at about 2.5 Ma. The migration of these hominoids out of Africa and among the different land masses had variable impacts on the local ecosystems, with the mass extinction of larger mammals in Australia and the Americas, generally coinciding with the arrival of *Homo sapiens*.

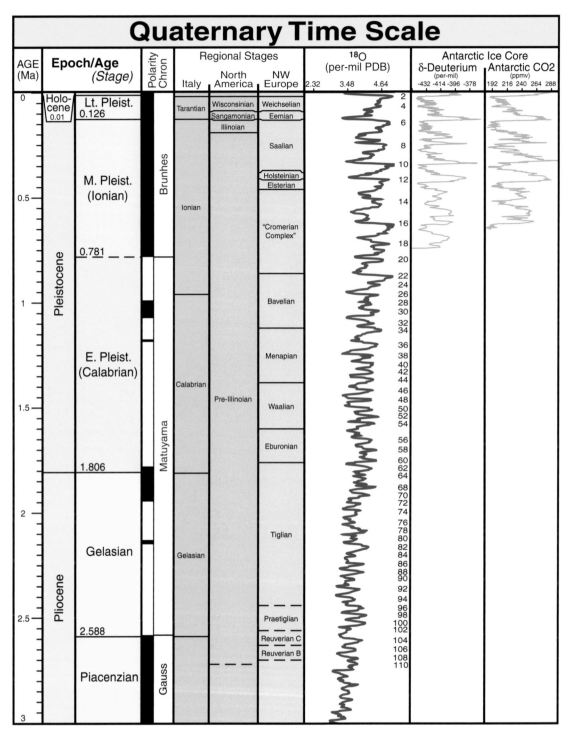

Figure 15.6. Numerical ages of epoch/series and age/stage boundaries and regional zonations for the latest Cenozoic with Antarctic ice-core records and ^{18}O curve. In this version, the Pleistocene begins at ~1.8 Ma (GSSP at Vrica, Italy) as established in 1983. Regional stages are from a chart prepared by Phil Gibbard *et al.* for the Quaternary Subcommission (see *www.quaternary.stratigraphy.org.uk*). The ^{18}O curve is from Crowhurst (2002). The δ-Deuterium curve is from Jouzel *et al.* (2004). Antarctic CO_2 curve was spliced together from ice-core databases archived at the NCDC Paleoclimatology Program (*www.ncdc.noaa.gov/paleo/icecore.html*): 0–11 ka = Taylor Dome (Indermühle *et al.*, 1999a); 11–27 ka = Taylor Dome (Smith *et al.*, 1999); 27–62 kyr = Taylor Dome (Indermühle *et al.*, 1999b). 64–417 kyr = Vostok (Barnola *et al.*, 2003); 417–649 kyr = Dome C (Siegenthaler *et al.*, 2005).

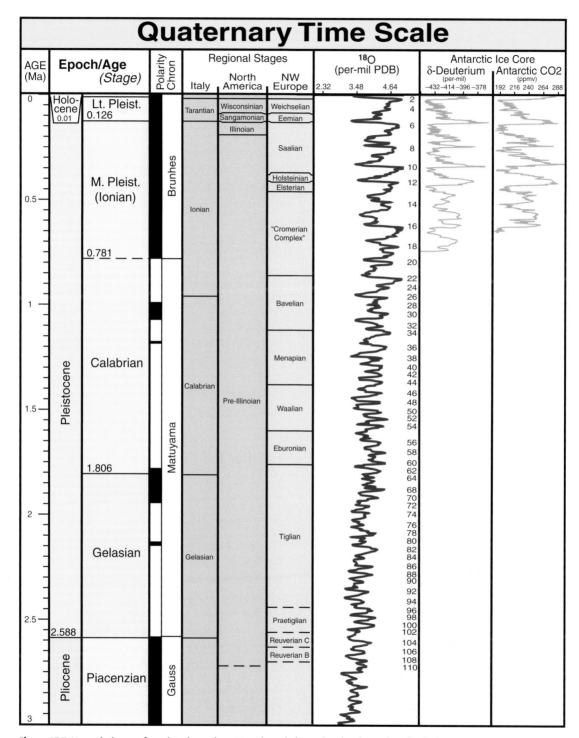

Figure 15.7. Numerical ages of epoch/series and age/stage boundaries and regional zonations for the latest Cenozoic with Antarctic ice-core records and ^{18}O curve. In this version, the Pleistocene (and Quaternary) encompasses the Gelasian Stage, as preferred by INQUA (see text).

Quaternary dating

In addition to methods of radiometric dating and Milankovitch cyclicity applied to older Cenozoic strata and artifacts, an arsenal of specialized dating techniques has been developed for Quaternary deposits. Accelerator mass spectrometers have enabled the range of carbon-14 dating to be extended to about 50 ka, although the calibration curve to calendar age is limited to about 25 ka. Cosmogenic nuclide dating is applied to landscape exposures and to sediment and skeleton burial to ~500 ka. Luminescence and electron spin resonance dating are applied to buried mineral grains and other materials.

Numerical time scale

Marine-isotope stages and magnetic polarity chrons are primarily dated using astronomical tuning in oceanic successions, either in cores or uplifted deposits. Indeed, the very detailed ice-core records of climate and atmospheric gases are dated in their older extent primarily by correlation to these marine-isotope stages.

In classical Quaternary stratigraphy, the record of glacial advances and deglaciation deposits established successions of named stages (e.g., Weichselian, Saalian and Elsterian for northwest Europe; Wisconsinian, Illinoian etc. for North America). However, these regional units and their records of terrestrial evolution are only partially correlated to the well-dated marine-isotope stages. Similarly, the fascinating record of human migrations and

tool-making developments awaits both the discovery of more regional artifacts and the ability to apply high-resolution dating methods to sites that lack interbedded volcanic ashes or precise correlations to magnetostratigraphy.

Acknowledgements

For further details/information, we recommend the "The Pleistocene and Holocene Epochs" by P. Gibbard and T. Van Kolfschoten and "The Neogene Period" by L. Lourens, F. Hilgen, N. J., Shackleton, J. Laskar, and D. Wilson (in *A Geologic Time Scale 2004*). Portions of the background material are from unpublished documents of the Neogene and Quaternary subcommissions.

Further reading

Balco, G., Rovey, C. W. II, and Stone, J. O. H., 2005. The first glacial maximum in North America. *Science*, 307: 222.

Barnola, J.-M., Raynaud, D., Lorius, C., and Barkov, N. I., 2003. Historical CO_2 record from the Vostok ice core. In: *Trends: A Compendium of Data on Global Change*. Oak Ridge, Tenn.: Carbon Dioxide Information Analysis Center, Oak Ridge National Laboratory, U.S. Department of Energy. Available on-line at http://cdiac.ornl.gov/trends/co2/vostok.htm

Bartoli, G., Sarnthein, M., Weinelt, M., Erlenkeuser, H., Garbe-Schonberg, D., and Lea, D. W., 2005. Final closure of Panama and the onset of northern hemisphere glaciation. *Earth and Planetary Science Letters*, **237**: 33–44.

Berggren, W. A., 2007. Status of the hierarchical subdivision of higher order marine Cenozoic chronostratigraphic units. *Stratigraphy*, **4**: 99–108.

Crowhurst, S. J., 2002. Composite isotope sequence. *The Delphi Project*. Available on-line at www.esc.cam.ac.uk/new/v10/research/institutes/godwin/body.html.

Crutzen, P. J., 2002. Geology of mankind. *Nature*, **415**: 23.

Elias, S. A. (ed.), 2007. *Encyclopedia of Quaternary Science* (4 vols). London: Elsevier.

Gibbard, P. L. and Pillans, B. (eds.) 2008. Special issue on the Quaternary, Episodes (in press).

Gibbard, P. L., 2003. Definition of the Middle–Upper Pleistocene boundary. *Global and Planetary Change*, **36**: 201–208.

Gibbard, P. L., Smith, A. G., Zalasiewicz, J. A., Barry, T. L., Cantrill, D., Coe, A. L., Cope, J. C. W., Gale, A. S., Gregory, F. J., Powell, J. H., Rawson, P. F., Stone, P., and Waters, C. N., 2005. What status for the Quaternary? *Boreas*, **34**: 1–6.

Haug, G. H., Ganopolski, A., Sigman, D. M., Rosell-Mele, A., Swann, G. E. A., Tiedemann, R., Jaccard, S. L., Bollmann, J., Masliln, M. A., Leng, M. J., and Eglinton, G., 2005. North Pacific seasonality and the glaciation of North America 2.7 million years ago. *Nature*, **33**: 821–825.

Head, M. J., and Gibbard, P. L. (eds.), 2005. *Early–Middle Pleistocene Transitions: The Land–Ocean Evidence*, Special Publication no. 247. London: Geological Society.

Indermühle A., Stocker, T. F., Joos, F., Fischer, H., Smith, H. J., Wahlen, M., Deck, B., Mastroianni, D., Tschumi, J., Blunier, T., Meyer, R., and Stauffer, B., 1999a. Holocene carbon-cycle dynamics based on CO_2 trapped in ice at Taylor Dome, Antarctica. *Nature*, **398**: 121–126.

Indermühle, A., Monnin, E., Stauffer, B., Stocker, T. F., and Wahlen, M., 1999b. Atmospheric CO_2 concentration from 60 to 20 kyr BP from the Taylor Dome ice core, Antarctica. *Geophysical Research Letters*, **27**: 735–738.

Jouzel, J., *et al.* 2004. *EPICA Dome C Ice Cores Deuterium Data*, IGBP PAGES/World Data Center for Paleoclimatology Data Contribution Series no. 2004-038. Boulder, Colo.: NOAA/NGDC Paleoclimatology Program.

Kerr, R. A., 2008. A time war over the period we live in. *Science*, **319**: 402–403.

Lowe, J. J., and Walker, M., 1997. *Reconstructing Quaternary Environments*, 2nd edn. New York: Prentice Hall.

Ogg, J., 2004. Introduction to concepts and proposed standardization of the term Quaternary. *Episodes*, **27**(2): 125–126.

Pillans, B., 2004. Proposal to redefine the Quaternary. *Episodes*, **27**(2): 127.

Siegenthaler, U., Stocker, T. F., Monnin, E., Lüthi, D., Schwander, J., and Stauffer, B.,

2005. *EPICA Dome C CO₂ Data 650 to 390 KYrBP*, IGBP PAGES/World Data Center for Paleoclimatology Data Contribution Series no. 2005–077. Boulder, Colo.: NOAA/NCDC Paleoclimatology Program.

Smith, H. J., Fischer, H., Mastroianni, D., Deck, B., and Wahlen, M., 1999. Dual modes of the carbon cycle since the Last Glacial Maximum. *Nature*, **400**: 248–250.

Walker, M., Johnsen, S., Rasmussen, S. O., Steffensen, J.-P., Popp, T., Gibbard, P., Hoek, W., Lowe, J., Andrews, J., Björck, S., Cwynar, L., Hughen, K., Kershaw, P., Kromer, B., Litt, T., Nakagawa, T., Newnham, R., and Schwander, J., 2008. A proposal for the Global Stratotype Section and Point (GSSP) and Global Standard Stratigraphic Age (GSSA) for the base of the Holocene Series/Epoch (Quaternary System/Period). *Episodes*, in press.

Walsh, S. L., 2008. The Neogene: origin, adoption, evolution, and controversy. *Earth Science Reviews*, **89**: 42–72.

Zalasiewicz, J., Williams, M., Smith, A., Barry, T. L., Coe, A. L., Bown, P. R., Brenchley, P., Cantrill, D., Gale, A., Gibbard, P., Gregory, F. J., Hounslow, M. W., Kerr, A. C., Pearson, P., Knox, R., Powell, J., Waters, C., Marshall, J., Oates, M., Rawson, R., and Stone, P., 2008. Are we now living in the Anthropocene? *GSA Today*, **18**(2): 4–8.

Selected on-line references

Quaternary Subcommission – *www.quaternary. stratigraphy.org.uk* – detailed inter-regional chart, status of Quaternary divisions, PDFs of major articles, and other information.

INQUA, the International Union for Quaternary Research (a full Science Union member of the International Council for Science) – *www. inqua.tcd.ie*

Definition and geochronologic/ chronostratigraphic rank of the term Quaternary, Recommendations by the Quaternary Task Group of ICS, IUGS and INQUA – *www.stratigraphy.org/Q2.pdf*

Wikipedia, especially Heinrich Events (contributed by GFDL) and Dansgaard – *en.wikipedia.org/wiki/Heinrich_event*, and *wiki/Dansgaard-Oeschger_event*

Authors

Philip Gibbard, Cambridge Quaternary, Department of Geography, University of Cambridge, Downing Place, Cambridge CB2 3EN, UK (Chair, Quaternary Subcommission of ICS)

Kim Cohen, Department of Physical Geography, Utrecht University, Heidelberglaan 2, Postbus 80.115, 3508 TC Utrecht, the Netherlands

James Ogg, Department of Earth and Atmospheric Sciences, Purdue University, 550 Stadium Mall Drive, West Lafayette, IN 47907, USA

Appendix 1

Standard colors of international divisions of geologic time

Figures A1 and A2 show the standard colors established by the *Commission for the Geological Map of the World* (CGMW/CCGM; *ccgm.free.fr*) for the international divisions of geologic time.

RGB Color Code according to the Commission for the Geological Map of the World (CGMW), Paris, France

The RGB color code is an additive model of Red, Green and Blue. Each is indicated on a scale from 0 (no pigment) to 255 (saturation of this pigment). "Devonian (203/140/205)" indicates a mixture of 203 Red, 140 Green and 205 Blue. The conversion from the reference CMYK values to the RGB codes utilizes Adobe® Illustrator® CS3's color function of "Emulate Adobe® Illustrator® 6.0" (menu Edit / Color Settings / Settings).

ATTENTION: For color conversions using a program other than Adobe® Illustrator®, it is necessary to conserve the reference CMYK, even if the resulting RGB values are slightly different.

* Definition of the Quaternary and revision of the Pleistocene are under discussion. Base of the Pleistocene is at 1.81 Ma (base of Calabrian), but may be extended to 2.59 Ma (base of Gelasian). The historic 'Tertiary' comprises the Paleogene and Neogene, and has no official rank.

Color composition by J.M. Pellé (BRGM, France)

Cenozoic / Mesozoic (upper Phanerozoic)

Phanerozoic 154/217/221 — Cenozoic 242/249/29 — Mesozoic 103/197/202

System	Series/Epoch	Stage
Quaternary* 249/249/127	Holocene 254/242/224	254/242/236
	Pleistocene 255/242/174	Upper 255/242/211; Middle 255/242/199; Lower 255/242/186; Gelasian 255/255/204
Neogene 255/230/25	Pliocene 255/255/153	Piacenzian 255/255/191; Zanclean 255/255/179
	Miocene 255/255/0	Messinian 255/255/115; Tortonian 255/255/102; Serravallian 255/255/89; Langhian 255/255/77; Burdigalian 255/255/65; Aquitanian 255/255/51
Paleogene 253/154/82	Oligocene 253/192/122	Chattian 254/230/170; Rupelian 254/217/154
	Eocene 253/180/108	Priabonian 253/205/161; Bartonian 253/192/145; Lutetian 252/180/130; Ypresian 252/167/115
	Paleocene 253/167/95	Thanetian 253/191/111; Selandian 254/191/101; Danian 253/180/98
Cretaceous 127/198/78	Upper 166/216/74	Maastrichtian 242/250/140; Campanian 230/244/127; Santonian 217/239/116; Coniacian 204/233/104; Turonian 191/227/93; Cenomanian 179/222/83
	Lower 140/205/87	Albian 204/234/151; Aptian 191/228/138; Barremian 179/223/127; Hauterivian 166/217/117; Valanginian 153/211/106; Berriasian 140/205/96

Mesozoic / Paleozoic (middle Phanerozoic)

Phanerozoic 154/217/221 — Mesozoic 103/197/202 — Paleozoic 153/192/141

Jurassic 52/178/201	Upper 179/227/238	Tithonian 217/241/247; Kimmeridgian 204/236/244; Oxfordian 191/231/241
	Middle 128/207/216	Callovian 191/231/229; Bathonian 179/226/227; Bajocian 166/221/224; Aalenian 154/217/221
	Lower 66/174/208	Toarcian 153/206/227; Pliensbachian 128/197/221; Sinemurian 103/188/216; Hettangian 78/179/211
Triassic 129/43/146	Upper 189/140/195	Rhaetian 227/185/219; Norian 214/170/211; Carnian 201/155/203
	Middle 177/104/177	Ladinian 201/131/191; Anisian 188/117/183
	Lower 152/57/153	Olenekian 176/81/165; Induan 164/70/159
Permian 240/64/40	Lopingian 251/167/148	Changhsingian 252/192/178; Wuchiapingian 252/180/162
	Guadalupian 251/116/92	Capitanian 251/154/133; Wordian 251/141/118; Roadian 251/128/105
	Cisuralian 239/88/69	Kungurian 227/135/118; Artinskian 227/123/104; Sakmarian 227/111/92; Asselian 227/99/80
Carboniferous 103/165/153	Pennsylvanian 153/194/181 (Upper 191/208/186; Middle 166/199/183; Lower 140/190/180)	Gzhelian 204/212/199; Kasimovian 191/208/197; Moscovian 199/203/185; Bashkirian 153/194/181
	Mississippian 103/143/102 (Upper 179/190/108... ; Middle 153/180/108; Lower 128/171/108)	Serpukhovian 191/194/107; Visean 166/185/108; Tournaisian 140/176/108

Paleozoic / Precambrian (lower Phanerozoic & Precambrian)

Phanerozoic 154/217/221 — Paleozoic 153/192/141

Devonian 203/140/55	Upper 241/225/157	Famennian 242/237/197; Frasnian 242/237/173
	Middle 241/200/104	Givetian 241/225/133; Eifelian 241/213/118
	Lower 229/172/77	Emsian 229/208/117; Pragian 229/196/104; Lochkovian 229/183/90
Silurian 179/225/182	Pridoli 230/245/225	230/245/225
	Ludlow 191/230/207	Ludfordian 217/240/223; Gorstian 204/236/221
	Wenlock 179/225/194	Homerian 204/235/209; Sheinwoodian 191/230/195
	Llandovery 153/215/179	Telychian 191/230/207; Aeronian 179/225/194; Rhuddanian 166/220/181
Ordovician 0/146/112	Upper 127/202/147	Hirnantian 166/219/171; Katian 153/214/158; Sandbian 140/208/148
	Middle 77/180/126	Darriwilian 116/198/156; Dapingian 102/192/146
	Lower 26/157/111	Floian 65/176/135; Tremadocian 51/169/126
Cambrian 127/160/86	Furongian 179/224/149	Stage 10 230/245/201; Stage 9 217/240/187; Paibian 204/235/174
	Series 3 166/207/134	Guzhangian 204/223/170; Drumian 191/217/157; Stage 5 179/212/146
	Series 2 153/192/120	Stage 4 179/202/142; Stage 3 166/197/131
	Terreneuvian 140/176/108	Stage 2 166/186/128; Fortunian 153/181/117

Precambrian 247/67/112

Eonothem/Erathem	System	Color
Proterozoic 247/53/99	Neoproterozoic 254/179/66	Ediacaran 254/217/106; Cryogenian 254/204/92; Tonian 254/191/78
	Mesoproterozoic 253/180/98	Stenian 254/217/154; Ectasian 253/204/138; Calymmian 253/192/122
	Paleoproterozoic 247/67/112	Statherian 248/117/167; Orosirian 247/104/152; Rhyacian 247/91/137; Siderian 247/79/124
Archean 240/4/127	Neoarchean 249/155/193	
	Mesoarchean 247/104/169	
	Paleoarchean 244/68/159	
	Eoarchean 218/3/127	
Hadean 174/2/126		

CMYK Color Code according to the Commission for the Geological Map of the World (CGMW), Paris, France

The CMYK color code is an additive model with percentages of Cyan, Magenta, Yellow and Black. For example: the CMYK color for Devonian (20/40/75/0) is a mixture of 20% Cyan, 40% Magenta, 75% Yellow and 0% Black. The CMYK values are the primary reference system for designating the official colors for these geological units.

* Definition of the Quaternary and revision of the Pleistocene are under discussion. Base of the Pleistocene is at 1.81 Ma (base of Calabrian), but may be extended to 2.59 Ma (base of Gelasian). The historic "Tertiary" comprises the Paleogene and Neogene, and has no official rank.

Color composition by J.M. Pellé (BRGM, France)

Precambrian (0/75/30/0)

Proterozoic (0/80/35/0)

Neoproterozoic (0/30/70/0)		Ediacaran (0/15/55/0)
		Cryogenian (0/20/60/0)
		Tonian (0/25/65/0)
Mesoproterozoic (0/30/55/0)		Stenian (0/15/35/0)
		Ectasian (0/20/40/0)
		Calymmian (0/25/45/0)
Paleoproterozoic (0/75/30/0)		Statherian (0/55/10/0)
		Orosirian (0/60/15/0)
		Rhyacian (0/65/20/0)
		Siderian (0/70/25/0)

Archean (0/100/0/0)

Era	Stage
Neoarchean (0/40/5/0)	(0/35/5/0)
Mesoarchean (0/60/5/0)	(0/50/5/0)
Paleoarchean (0/75/0/0)	(0/60/0/0)
Eoarchean (10/100/0/0)	(5/90/0/0)

Hadean (30/100/0/0)

Phanerozoic (40/0/5/0) — Paleozoic (40/10/40/0)

Devonian (20/40/75/0)
Upper (5/10/35/0)	Famennian (5/5/20/0)
	Frasnian (5/5/30/0)
Middle (5/20/55/0)	Givetian (5/10/45/0)
	Eifelian (5/15/50/0)
Lower (10/30/65/0)	Emsian (10/15/50/0)
	Pragian (10/20/55/0)
	Lochkovian (10/25/60/0)

Silurian (30/0/25/0)
Pridoli (10/0/10/0)	(10/0/10/0)
Ludlow (25/0/15/0)	Ludfordian (15/0/10/0)
	Gorstian (20/0/10/0)
Wenlock (30/0/20/0)	Homerian (20/0/15/0)
	Sheinwoodian (25/0/20/0)
Llandovery (40/0/25/0)	Telychian (25/0/15/0)
	Aeronian (30/0/20/0)
	Rhuddanian (35/0/25/0)

Ordovician (100/0/60/0)
Upper (50/0/40/0)	Hirnantian (35/0/30/0)
	Katian (40/0/35/0)
	Sandbian (45/0/40/0)
Middle (70/0/50/0)	Darriwilian (55/0/35/0)
	Dapingian (60/0/40/0)
Lower (90/0/60/0)	Floian (75/0/45/0)
	Tremadocian (80/0/50/0)

Cambrian (50/20/65/0)
Furongian (30/0/40/0)	Stage 10 (10/0/20/0)
	Stage 9 (15/0/25/0)
	Paibian (20/0/30/0)
Series 3 (35/5/45/0)	Guzhangian (20/5/30/0)
	Drumian (25/5/35/0)
	Stage 5 (30/5/40/0)
Series 2 (40/10/50/0)	Stage 4 (30/10/40/0)
	Stage 3 (35/10/45/0)
	Stage 2 (35/15/45/0)
Terreneuvian (45/15/55/0)	Fortunian (40/15/50/0)

Phanerozoic (40/0/5/0) — Mesozoic (60/0/10/0)

Jurassic (80/0/5/0)
Upper (30/0/0/0)	Tithonian (15/0/0/0)
	Kimmeridgian (20/0/0/0)
	Oxfordian (25/0/5/0)
Middle (50/0/5/0)	Callovian (25/0/5/0)
	Bathonian (30/0/5/0)
	Bajocian (35/0/5/0)
	Aalenian (40/0/5/0)
Lower (75/5/0/0)	Toarcian (40/5/0/0)
	Pliensbachian (50/5/0/0)
	Sinemurian (60/5/0/0)
	Hettangian (70/5/0/0)

Triassic (50/80/0/0)
Upper (25/40/0/0)	Rhaetian (10/25/0/0)
	Norian (15/30/0/0)
	Carnian (20/35/0/0)
Middle (30/55/0/0)	Ladinian (20/45/0/0)
	Anisian (25/50/0/0)
Lower (40/75/0/0)	Olenekian (30/65/0/0)
	Induan (35/70/0/0)

Phanerozoic (40/0/5/0) — Paleozoic (40/10/40/0)

Permian (5/75/75/0)
Lopingian (0/35/30/0)	Changhsingian (0/25/20/0)
	Wuchiapingian (0/30/25/0)
Guadalupian (0/55/50/0)	Capitanian (0/40/35/0)
	Wordian (0/45/40/0)
	Roadian (0/50/45/0)
Cisuralian (5/65/60/0)	Kungurian (10/45/40/0)
	Artinskian (10/50/45/0)
	Sakmarian (10/55/50/0)
	Asselian (10/60/55/0)

Carboniferous (60/15/30/0)
Subsystem	Series	Stage
Pennsylvanian (0/10/20/0)	Upper (25/10/20/0)	Gzhelian (20/10/15/0)
		Kasimovian (25/10/15/0)
	Middle (35/10/20/0)	Moscovian (30/10/20/0)
	Lower (45/10/20/0)	Bashkirian (40/10/20/0)
Mississippian (0/25/55/0)	Upper (30/15/55/0)	Serpukhovian (25/15/55/0)
	Middle (40/15/55/0)	Visean (35/15/55/0)
	Lower (60/15/55/0)	Tournaisian (45/15/55/0)

Phanerozoic (40/0/5/0) — Cenozoic (5/0/90/0)

Quaternary* (0/0/50/0)
Holocene (0/5/10/0)	(0/5/5/0)
Pleistocene (0/5/30/0)	Upper (0/5/15/0)
	Middle (0/5/20/0)
	Lower (0/5/25/0)

Neogene (0/10/90/0)
Pliocene (0/0/40/0)	Gelasian (0/0/20/0)
	Piacenzian (0/0/25/0)
	Zanclean (0/0/30/0)
Miocene (0/0/100/0)	Messinian (0/0/55/0)
	Tortonian (0/0/60/0)
	Serravallian (0/0/65/0)
	Langhian (0/0/70/0)
	Burdigalian (0/0/75/0)
	Aquitanian (0/0/80/0)

Paleogene (0/40/60/0)
Oligocene (0/25/45/0)	Chattian (0/10/30/0)
	Rupelian (0/15/35/0)
Eocene (0/30/50/0)	Priabonian (0/20/30/0)
	Bartonian (0/25/35/0)
	Lutetian (0/30/40/0)
	Ypresian (0/35/45/0)
Paleocene (0/35/55/0)	Thanetian (0/25/50/0)
	Selandian (0/25/55/0)
	Danian (0/30/55/0)

Phanerozoic (40/0/5/0) — Mesozoic (60/0/10/0)

Cretaceous (50/0/75/0)
Upper (35/0/75/0)	Maastrichtian (5/0/45/0)
	Campanian (10/10/50/0)
	Santonian (15/0/55/0)
	Coniacian (20/0/60/0)
	Turonian (25/0/65/0)
	Cenomanian (30/0/70/0)
Lower (45/0/70/0)	Albian (20/0/40/0)
	Aptian (25/0/45/0)
	Barremian (30/0/50/0)
	Hauterivian (35/0/55/0)
	Valanginian (40/0/60/0)
	Berriasian (45/0/65/0)

Appendix 2

Ratified GSSPs for geologic stages

Cenozoic

Neogene and Quaternary

Holocene Series

Walker, M., Johnsen, S., Rasmussen, S. O., Steffensen, J.-P., Popp, T., Gibbard, P., Hoek, W., Lowe, J., Andrews, J., Björck, S., Cwynar, L., Hughen, K., Kershaw, P., Kromer, B., Litt, T., Nakagawa, T., Newnham, R., Schwander, J., 2008. A proposal for the Global Stratotype Section and Point (GSSP) and Global Standard Stratigraphic Age (GSSA) for the base of the Holocene Series/Epoch (Quaternary System/Period). *Journal of Quaternary Science*, submitted.

Pleistocene Series (*proposed Calabrian Stage*)

Aguirre, E., and Pasini, G., 1985. The Pliocene–Pleistocene Boundary. *Episodes*, 8(2): 116–120.

Bassett, M. G., 1985. Towards a "common language" in stratigraphy. *Episodes*, 8(2): 87.

Gelasian (*potential base of Pleistocene Series and Quaternary System*)

Rio, D., Sprovieri, R., Castradori, D., and Di Stefano, E., 1998. The Gelasian Stage (Upper Pliocene): a new unit of the global standard chronostratigraphic scale. *Episodes*, **21**(2): 82–87.

Piacenzian

Castradori, D., Rio, D., Hilgen, F. J., and Lourens, L. J., 1998. The Global Standard Stratotype-section and Point (GSSP) of the Piacenzian Stage (Middle Pliocene). *Episodes*, **21** (2): 88–93.

Zanclean (*base of Pliocene Series*)

Van Couvering, J. A., Castradori, D., Cita, M. B., Hilgen, F. J., and Rio, D., 2000. The base of the Zanclean Stage and of the Pliocene Series. *Episodes*, **23**(3): 179–187.

Messinian

Hilgen, F. J., Iaccarino, S., Krijgsman, W., Villa, G., Langereis, C. G., and Zachariasse, W. J., 2000. The Global Boundary Stratotype Section and Point (GSSP) of the Messinian Stage (uppermost Miocene). *Episodes*, **23**(3): 172–178.

Tortonian

Hilgen, F. J., Abdul Aziz, H., Bice, D., Iaccarino, S., Krijgsman, W., Kuiper, K., Montanari, A., Raffi, I., Turco, E., and Zachariasse, W. J., 2005. The Global Boundary Stratotype Section and Point (GSSP) of the Tortonian Stage (Upper Miocene) at Monte dei Corvi. *Episodes*, **28**(1): 6–17.

Aquitanian (*base of Miocene Series and Neogene System*)

Steiniger, F. F., Aubry, M. P., Berggren, W. A., Biolzi, M., Borsetti, A. M., Cartlidge, J. E., Cati, F., Corfield, R., Gelati, R., Iaccarino, S., Napoleone, C., Ottner, F., Roegl, F., Roetzel, R., Spezzaferri, S., Tateo, F., Villa, G., and Zevenboom, D., 1997. The Global Stratotype Section and Point (GSSP) for the Base of the Neogene. *Episodes*, **20**(1): 23–28.

Paleogene

Rupelian (*base of Oligocene Series*)

Primoli Silva, I., and Jenkins, D. G., 1993. Decision on the Eocene–Oligocene boundary stratotype. *Episodes*, **16**(3): 379–382.

Ypresian (*base of Eocene Series*)

Aubry, M., Ouda, K., Dupuis, C., Berggren, W. A., Van Couvering, J. A., and the Members of the Working Group on the Paleocene/Eocene Boundary, 2007. The Global Standard Stratotype-section and Point (GSSP) for the base of the Eocene Series in the Dababiya section (Egypt). *Episodes*, **30**(4): 271–286.

Dupuis, C., Aubry, M., Steurbaut, E., Berggren, W. A., Ouda, K., Magioncalda, R., Cramer, B. S., Kent, D. V., Speijer, R. P., and Heilmann-Clausen, C., 2003. The Dababiya Quarry Section: lithostratigraphy, clay mineralogy, geochemistry and paleontology. *Micropaleontology*, **49**(1): 41–59.

Danian (*base of Paleocene Series and Paleogene System*)

Molina, E., Alegret, L., Arenillas, I., Arz, J. A., Gallala, N., Hardenbol, J., von Salis, K., Steurbaut, E., Vandenberghe, N., and Zaghbib-Turki, D. 2006. The Global Boundary Stratotype Section and Point for the base of the Danian Stage (Paleocene, Paleogene, "Tertiary," Cenozoic) at El Kef, Tunisia: original definition and revision. *Episodes*, **29**(4): 263–278.

Mesozoic

Cretaceous

Maastrichtian

Odin, G. S. (ed.), 2001. *The Campanian–Maastrichtian Boundary: Characterisation at*

Tercis les Bains (France) and Correlation with Europe and other Continents, IUGS Special Publication Series vol. 36. Amsterdam: Elsevier.

Odin, G. S., and Lamaurelle, M. A., 2001. The global Campanian–Maastrichtian stage boundary. *Episodes*, **24**(4): 229–238.

Turonian

Kennedy, W. J., Walaszcyk, I., and Cobban, W. A., 2005. The Global Boundary Stratotype Section and Point for the base of the Turonian Stage of the Cretaceous: Pueblo, Colorado, USA. *Episodes*, **28**(2): 93–104.

Cenomanian

Kennedy, W. J., Gale, A. S., Lees, J. A., and Caron, M., 2004. The Global Boundary Stratotype Section and Point (GSSP) for the base of the Cenomanian Stage, Mont Risou, Hautes-Alpes, France. *Episodes*, **27**(1): 21–32.

Jurassic

Bajocian

Pavia, G., and Enay, R., 1997. Definiton of the Aalenian–Bajocian Stage Boundary. *Episodes*, **20**(1): 16–22.

Aalenian

Cresta, S., Goy, A., Ureta, S., Arias, C., Barrón, E., Bernard, J., Canales, M. L., García-Joral, F., García-Romero, E., Gialanella, P. R., Gomes, J. J., González, J. A., Herrero, C., Marínez, G.,

Osete, M. L., Perilli, N., and Villalaín, J. J., 2001. The Global Boundary Stratotype Section and Point (GSSP) of the Toarcian–Aalenian Boundary (Lower–Middle Jurassic). *Episodes*, **24**(3): 166–175.

Pliensbachian

Meister, C., Aberhan, M., Blau, J., Dommergues, J.-L., Feist-Burkhardt, S., Hailwood, E. A., Hart, M., Hesselbo, S. P., Hounslow, M. W., Hylton, M., Morton, N., Page, K., and Price, G. D., 2006. The Global Boundary Stratotype Section and Point (GSSP) for the base of the Pliensbachian Stage (Lower Jurassic), Wine Haven, Yorkshire, UK. *Episodes*, **29**(2): 93–106.

Sinemurian

Bloos, G., and Page, K. N., 2002. The Global Stratotype Section and Point for base of the Sinemurian Stage (Lower Jurassic). *Episodes*, **25**(1): 22–28.

Triassic

Carnian

Mietto, P., Andreetta, R., Broglio Loriga, C., Buratti, N., Cirilli, S., De Zanche, V., Furin, S., Gianolla, P., Manfrin, S., Muttoni, G., Neri, C., Nicora, A., Posenato, R., Preto, N., Rigo, M., Roghi, G., and Spötl, C., 2007. A candidate of the Global Boundary Stratotype Section and Point for the base of the Carnian Stage (Upper Triassic): GSSP at the base of the *canadensis* Subzone (FAD of *Daxatina*) in the Prati di

Stuores/Stuores Wiesen section (Southern Alps, NE Italy). *Albertiana*, **36**: 78–97. [*Proposal accepted by ICS, pending IUGS ratification*]

Ladinian

Brack, P., Rieber, H., Nicora, A., and Mundil, R., 2005. The Global Boundary Stratotype Section and Point (GSSP) of the Ladinian Stage (Middle Triassic) at Bagolino (Southern Alps, Northern Italy) and its implications for the Triassic time scale. *Episodes*, **28**(4): 233–244.

Induan (*base of Triassic System*)

Yin, H., Zhang, K., Tong, J., Yang, Z., and Wu, S., 2001. The Global Stratotype Section and Point (GSSP) of the Permian–Triassic Boundary. *Episodes*, **24**(2): 102–114.

Paleozoic

Permian

Changhsingian

Jin, Y., Wang, Y., Henderson, C., Wardlaw, B. R., Shen, S., and Cao, C., 2006. The Global Boundary Stratotype Section and Point (GSSP) for the base of Changhsingian Stage (Upper Permian). *Episodes*, **29**(3): 175–182.

Wuchiapingian

Ji, Y., Shen, S., Henderson, C. M., Wang, X., Wang, W., Wang, Y., Cao, C., and Shang, Q., 2006. The Global Stratotype Section and Point (GSSP) for the boundary between the Capitanian and Wuchiapingian Stage (Permian). *Episodes*, **29**(4): 253–262.

Captianian, Wordian, Roadian

Details of this set of ratified GSSPs are being submitted to Episodes (*by Wardlaw* et al.) *as of May 2008.*

Asselian (*base of Permian System*)

Davydov, V. I., Glenister, B. F., Spinosa, C., Ritter, S. M., Chemykh, V. V., Wardlaw, B. R., and Snyder, W. S., 1998. Proposal of Aidaralash as Global Stratotype Section and Point (GSSP) for base of the Permian System. *Episodes*, **21**(1): 11–18.

Carboniferous

Bashkirian

Lane, H. R., Brenckle, P. L., Baesemann, J. F., and Richards, B., 1999. The IUGS boundary in the middle of the Carboniferous: Arrow Canyon, Nevada, USA. *Episodes*, **22**(4): 272–283.

Richards, B. C., Lane, H. R., and Brenckle, P. L., 2002. The IUGS Mid-Carboniferous (Mississippian–Pennsylvanian) global boundary stratotype section and point at Arrow Canyon, Nevada, USA. In: *Carboniferous and Permian of the World*, eds. L. V. Hills, C. M. Henderson, and E. M. Bamber. *Memoir Canadian Society of Petroleum Geologists*, **19**: 802–831.

Visean

Devuyst, F. X., Hance, L., Hou, H., Wu, X., Tian, S., Coen, M., and Sevastopulo, G. 2003. A proposed Global Stratotype Section and Point for the base of the Visean Stage (Carboniferous): the Pengchong section, Guangxi, South China. *Episodes*, **26**(2): 105–115. [*A enhanced version of this proposal was accepted by ICS. Ratification announcement is being submitted to* Episodes *(by Devuyst* et al.*) as of May 2008.*]

Tournaisian (*base of Carboniferous System*)

Paproth, E., Feist, R., and Flaijs, G., 1991. Decision on the Devonian–Carboniferous boundary stratotype. *Episodes*, **14**(4): 331–336.

Devonian

Famennian

Klapper, G., Feist, R. Becker, R. T., and House, M. R., 1993. Definition of the Frasnian/Famennian Stage boundary. *Episodes*, **16**(4): 433–441.

Frasnian

Klapper, G., Feist, R., and House, M. R., 1987. Decision on the Boundary Stratotype for the Middle/Upper Devonian Series Boundary. *Episodes*, **10**(2): 97–101.

Givetian

Walliser, O. H., Bultynck, P., Weddige, K., Becker, R. T., and House, M. R., 1995.

Definition of the Eifelian–Givetian Stage Boundary. *Episodes*, **18**(3): 107–115.

Eifelian

Ziegler, W., and Klapper, G., 1985. Stages of the Devonian System. *Episodes*, **8**(2): 104–109.

Emsian

Yolkin, E. A., Kim, A. I., Weddige, K., Talent, J. A., and House, M. R., 1997. Definition of the Pragian/Emsian Stage boundary. *Episodes*, **20**(4): 235–240.

Pragian

Chlupác, I., and Oliver, W. A., 1989. Decision on the Lochkovian–Pragian Boundary Stratotype (Lower Devonian). *Episodes*, **12**(2): 109–113.

Lochkovian (*base of Devonian System*)

Chlupác, I., and Kukal, Z., 1977. The boundary stratotype at Klonk. In: *The Silurian–Devonian Boundary*, ed A. Martinsson. *International Union of Geological Sciences, Series A*, **5**: 96–109.

Silurian

Prídolí Series

Holland, C. H., 1985. Series and stages of the Silurian System. *Episodes*, **8**(2): 101–103.

Kríz, J., 1989. The Prídolí Series in the Prague Basin (Barrandium area, Bohemia). In:

A Global Standard for the Silurian System, eds. C. H. Holland and M. G. Bassett. *Geological Series, National Museum of Wales*, **9**: 90–100.

Martinsson, A., Bassett, M. G., and Holland, C. H., 1981. Ratification of standard chronostratigraphic divisions and stratotypes for the Silurian System. *Lethaia*, **14**: 168.

Ludfordian

Holland, C. H., 1985. Series and stages of the Silurian System. *Episodes*, 8(2): 101–103.

Lawson, J. D., and White, D. E., 1989. The Ludlow Series in the Ludlow area. In: *A Global Standard for the Silurian System*, eds. C. H. Holland and M. G. Bassett. *Geological Series, National Museum of Wales*, **9**: 73–90.

Martinsson, A., Bassett, M. G., and Holland, C. H., 1981. Ratification of standard chronostratigraphic divisions and stratotypes for the Silurian System. *Lethaia*, **14**: 168.

Gorstian

Holland, C. H., 1985. Series and stages of the Silurian System. *Episodes*, 8(2): 101–103.

Lawson, J. D., and White, D. E., 1989. The Ludlow Series in the Ludlow area. In: *A Global Standard for the Silurian System*, eds. C. H. Holland and M. G. Bassett. *Geological Series, National Museum of Wales*, **9**: 73–90.

Martinsson, A., Bassett, M. G., and Holland, C. H., 1981. Ratification of standard chronostratigraphic divisions and stratotypes for the Silurian System. *Lethaia*, **14**: 168.

Homerian

Bassett, M. G., 1989. The Wenlock Series in the Wenlock area. In: *A Global Standard for the Silurian System*, eds. C. H. Holland and M. G. Bassett. *Geological Series, National Museum of Wales*, **9**: 51–73.

Holland, C. H., 1985. Series and stages of the Silurian System. *Episodes*, 8(2): 101–103.

Martinsson, A., Bassett, M. G., and Holland, C. H., 1981. Ratification of standard chronostratigraphic divisions and stratotypes for the Silurian System. *Lethaia*, **14**: 168.

Sheinwoodian

Bassett, M. G., 1989. The Wenlock Series in the Wenlock area. In: *A Global Standard for the Silurian System*, eds. C. H. Holland and M. G. Bassett. *Geological Series, National Museum of Wales*, **9**: 51–73.

Holland, C. H., 1985. Series and stages of the Silurian System. *Episodes*, 8(2): 101–103.

Martinsson, A., Bassett, M. G., and Holland, C. H., 1981. Ratification of standard chronostratigraphic divisions and stratotypes for the Silurian System. *Lethaia*, **14**: 168.

Telychian

Cocks, L. R. M., 1989. The Llandovery Series in the Llandovery area. In: *A Global Standard for the Silurian System*, eds. C. H. Holland and M. G. Bassett. *Geological Series, National Museum of Wales*, **9**: 36–50.

Holland, C. H., 1985. Series and stages of the Silurian System. *Episodes*, 8(2): 101–103.

Martinsson, A., Bassett, M. G., and Holland, C. H., 1981. Ratification of standard chronostratigraphic divisions and stratotypes for the Silurian System. *Lethaia*, 14: 168.

Aeronian

Cocks, L. R. M., 1989. The Llandovery Series in the Llandovery area. In: *A Global Standard for the Silurian System*, eds. C. H. Holland and M. G. Bassett. *Geological Series, National Museum of Wales* (Cardiff), 9: 36–50.

Holland, C. H., 1985. Series and stages of the Silurian System. *Episodes*, 8(2): 101–103.

Martinsson, A., Bassett, M. G., and Holland, C. H., 1981. Ratification of standard chronostratigraphic divisions and stratotypes for the Silurian System. *Lethaia*, 14: 168.

Rhuddanian (*base of Silurian System*)

Cocks, L. R. M., 1985. The Ordovician–Silurian Boundary. *Episodes*, 8(2): 98–100.

Ordovician

[Official nomenclature of the stages Sandbian, Dapingian, and Floian had not yet been published as of May, 2008.]

Hirnantian

Chen, X., Rong, J., Fan, J., Zhan, R., Mitchell, C. E., Harper, D. A. T., Melchin, M. J.,

Peng, P., Finney, S. C., and Wang, X., 2006. The Global Boundary Stratotype Section and Point (GSSP) for the base of the Hirnantian Stage (the uppermost of the Ordovician System). *Episodes*, 29(3): 183–196.

Katian

Goldman, D., Leslie, S. A., Nolvak, J., Young, S., Bergström, S. M., and Huff, W. D., 2007. The Global Stratotype Section and Point (GSSP) for the base of the Katian Stage of the Upper Ordovician Series at Black Knob Ridge, Southeastern Oklahoma, USA. *Episodes*, 30(4): 258–270.

Sandbian

Bergström, S. M., Finney, S. C., Chen, X., Pålsson, C., Wang, Z., and Grahn, Y., 2000. A proposed global boundary stratotype for the base of the Upper Series of the Ordovician System: the Fågelsång section, Scania, southern Sweden. *Episodes*, 23(3): 102–109.

Darriwilian

Mitchell, C. E., Chen, X., Bergström, S. M., Zhang, Y., Wang, Z., Webby, B. D., and Finney, S. C., 1997. Definition of a global stratotype for the Darriwilian Stage of the Ordovician System. *Episodes*, 20(3): 158–166.

Dapingian

Wang, X., Stouge, S., Erdtmann, B., Chen, X., Li, Z., Wang, C., Zeng, Q., Zhou, Z., and

Chen, H., 2005. A proposed GSSP for the base of the Middle Ordovician series: the Huanghuachang Section, Yichang, China. *Episodes*, **28**(2): 105–117. [*Proposal that was accepted by ICS*]

Wang, X., Stouge, S., Erdtmann, B., Chen, X., Li, Z., Wang, C., Finney, S. C., Zeng, Q., Zhou, Z., and Chen, H., 2008. The Global Stratotype Section and Point for the base of the Middle Ordovician Series and the Third Stage. [*Publication pending in* Episodes *as of May, 2008*]

Floian

Bergström, S. M., Löfgren, A., and Maletz, J., 2004. The GSSP of the Second (Upper) Stage of the Lower Ordovician Series: Diabasbrottet at Hunneberg, Province of Västergötland, Southwestern Sweden. *Episodes*, **27**(4): 265–272.

Tremadocian (*base of Ordovician System*)

Cooper, R. A., Nowlan, G. S., and Williams, S. H., 2001. Global Stratotype Section and Point for base of the Ordovician System. *Episodes*, **24**(1): 19–28.

Cambrian

Paibian

Peng, S., Babcock, L. E., Robison, R. A., Lin, H., Rees, M. N., and Saltzman, M. R., 2004. Global Standard Stratotype-section and Point (GSSP) of the Furongian Series and Paibian Stage (Cambrian). *Lethaia*, **37**: 365–379.

Drumian

Babcock, L. E., Robison, R. A., Rees, M. N., Peng, S., and Saltzman, M. R., 2007. The Global boundary Stratotype Section and Point (GSSP) of the Drumian Stage (Cambrian) in the Drum Mountains, Utah, USA. *Episodes*, **30**(2): 85–95.

Fortunian (*base of Cambrian System*)

Brasier, M., Cowie, J., and Taylor, M., 1994. Decision on the Precambrian–Cambrian boundary stratotype. *Episodes*, **17**(1/2): 95–100.

Landing, E., Peng, S., Babcock, L. E., Geyer, G., and Moczydlowska-Vidal, M., 2007. Global standard names for the lowermost Cambrian Series and Stage. *Episodes*, **30**(4): 287–289.

Precambrian Units

Ediacaran

Knoll, A. H., Walter, M. R., Narbonne, G. M., and Christie-Blick, N., 2006. The Ediacaran Period: a new addition to the geologic time scale. *Lethaia*, **39**: 13–30.

Precambrian periods, eras, and eons

Plumb, K. A., and James, H. L., 1986. Subdivision of Precambrian time: recommendations and suggestions by the Subcommission on Precambrian Stratigraphy. *Precambrian Research*, **32**: 65–92.

Plumb, K. A., 1991. New Precambrian time scale. *Episodes*, **14**(2): 139–140.

Index